AF335289

STRUCTURAL PERFORMANCE OF FLEXIBLE PIPES

STRUCTURAL PERFORMANCE OF FLEXIBLE PIPES

PROCEEDINGS OF THE FIRST NATIONAL CONFERENCE ON FLEXIBLE PIPES
COLUMBUS / OHIO / 21-23 OCTOBER 1990

Structural Performance of Flexible Pipes

Edited by
SHAD M. SARGAND & GAYLE F. MITCHELL
Center for Geotechnical and Groundwater Research, Ohio University, Athens, Ohio, USA

JOHN OWEN HURD
Ohio Department of Transportation, Columbus, Ohio, USA

Sponsored by the Center for Geotechnical and Groundwater Research, in cooperation with the Ohio Department of Transportation

A.A.BALKEMA / ROTTERDAM / BROOKFIELD / 1990

PAPER REVIEW COMMITTEE

Shad M. Sargand, Russ Professor, Civil Engineering, Ohio University
George Abdel-Sayed, Professor, Civil Engineering, University of Windsor, Canada
Joseph V. Benak, Professor, Civil Engineering, University of Nebraska
David R. Bowman, Geotechnical Engineer, BBC&M Engineering, Inc.
Lester H. Gabriel, Professor, Civil Engineering, California State University, Sacramento
James B. Goddard, Chief Marketing Engineer, Advanced Drainage Systems, Inc.
Glenn Hazen, Professor, Civil Engineering, Ohio University
John Hurd, Hydraulic Research Engineer, Ohio Department of Transportation
W.G. Krouse, Bridge Engineer, Ohio Department of Transportation
Robert Landreth, Chief, Landfill Technology, US Environmental Protection Agency
Gayle F. Mitchell, Director, CGGR, Civil Engineering, Ohio University
Samuel C. Musser, Division Engineer, Syro Steel Company
Alan Rauch, Research Engineer, Applied Research Associates
Raymond B. Seed, Associate Professor, Civil Engineering, University of California, Berkeley
Ernest T. Selig, Professor, Civil Engineering, University of Massachusetts

SUPPORTING ORGANIZATIONS

Advanced Drainage Systems, Inc.
Syro Steel Company
Lane Enterprises, Inc.
Hancor, Inc.
College of Engineering and Technology, Ohio University
American Society of Civil Engineers, Central Ohio Section

The texts of the various papers in this volume were set individually by typists under the supervision of each of the authors concerned.

Authorization to photocopy items for internal or personal use, or the internal or personal use of specific clients, is granted by A.A. Balkema, Rotterdam, provided that the base fee of US$1.00 per copy, plus US$0.10 per page is paid directly to Copyright Clearance Center, 27 Congress Street, Salem, MA 01970, USA. For those organizations that have been granted a photocopy license by CCC, a separate system of payment has been arranged. The fee code for users of the Transactional Reporting Service is. 90 6191 165 6/90 US$1.00 ı US$0.10.

Published by
A.A. Balkema, P.O. Box 1675, 3000 BR Rotterdam, Netherlands
A.A. Balkema Publishers, Old Post Road, Brookfield, VT 05036, USA

ISBN 90 6191 165 6

© 1990 A.A. Balkema, Rotterdam
Printed in the Netherlands

Structural Performance of Flexible Pipes, Sargand, Mitchell & Hurd (eds) © 1990 Balkema, Rotterdam. ISBN 90 6191 165 6

Table of contents

Structural Performance of Flexible Pipes, Sargand, Mitchell & Hurd (eds) © 1990 Balkema, Rotterdam. ISBN 90 6191 165 6

Preface

In recent years, as a direct result of the broadening and expanding application of flexible pipes in infrastructure systems, intensive research on the structural performance, design and analysis, and modeling of these structures has been conducted. Most of the information related to structural design of flexible pipes has been provided by manufacturers, academia and government; however, this information often has not been collected into a comprehensive unit. This publication provides the forum for integrating these different viewpoints on structural performance and other elements of flexible pipes.

The papers in this publication include manuscripts on metal pipes, plastic pipes, design and analysis, and those of a general nature. The lead paper is the keynote address delivered by Dr Lester H.Gabriel, Professor of Civil Engineering, California State University, Sacramento. The reviewing procedure included review of both the submitted abstracts and papers by the Review Committee. Their objective was to assure clear and reasonable presentation, scientific quality and satisfactory expression of the English language. The thoughtful and thorough review provided by the committee is greatly appreciated.

We would like to acknowledge individuals from the Ohio Department of Transportation (ODOT), who helped contribute to the success of the conference and thus the proceedings. Mr William Edwards, Engineer of Research and Development, aptly served as co-chair of the Organizing Committee and provided constructive suggestions that enhanced the quality of the conference. The support provided by Mr Kenneth M.Miller, Executive Assistant to the Director, and Mr John D.Herl, Hydraulic Engineer, contributed greatly to the success of the conference.

The keynote speakers for the conference, Dr J.Michael Duncan, University Distinguished Professor of Civil Engineering, Virginia Polytechnic Institute and State University, Dr Ernest T.Selig, Professor of Civil Engineering, University of Massachusetts, and Dr Lester H.Gabriel are nationally recognized authorities in the area of flexible pipes. They provided an overview on the structural performance of flexible pipes, which set an excellent tone for the conference. Acknowledgement is also due Dr William Marcuson, Chief, Geotechnical Laboratory, US Army Corps of Engineers, Vicksburg, who graciously delivered the luncheon address.

Another individual that should be recognized is Mr James Goddard, Chief Marketing Engineer, Advanced Drainage Systems, Inc., for his help in the organization of the conference and exhibits.

Appreciation is also expressed to Ms Helen Malone, secretary in the Center for Geotechnical and Groundwater Research (CGGR), for her assistance in the many details and expert typing required for a conference and proceedings of this type.

Shad M.Sargand
Gayle F.Mitchell

Structural Performance of Flexible Pipes, Sargand, Mitchell & Hurd (eds) © 1990 Balkema, Rotterdam. ISBN 90 6191 165 6

Keynote address: Pipe deflections – A redeemable asset

Lester H.Gabriel
California State University, Sacramento, Calif., USA

The adjective, redeemable, which is used in the title of this address, is intended to reflect the sense of opportunity for release from blame for a wrong that has been committed. It is often the case, however, that a wrong is a wrong only because it is so defined. But if indeed the alleged wrong is improperly cast, the sense of a release from blame should be a doubly welcome event.

It is my intention to show that the deflections of pipes under load is the asset, or resource, to be considered an attribute which enhances the structural performance of flexible pipes. Pipe deformation, a benefactor of all loaded pipes, rigid and flexible, adds surety to the service performance of a pipe or culvert. Unfortunately, it has to suffer the indignity of being the object of redemption – just because it has been falsely accused as a liability and often does not receive the benefit of a fair trial. The juries have often been biased – the judges have often been intimidated.

It would be too presumptuous of me to believe that I personally, have the power to right a wrong with a telling of a truth. I am satisfied to only lend my voice to those who also know that, without pipe deflections, the forces and moments internal to a pipe could be unbearable and disaster could be a likely outcome.

But there is a truth. And the truth is that a compliant structure, one capable of responding with deformations, is a structure that will convert otherwise excesses of stresses, from one mode of resistance to imposed load, to another mode of resistance to the same load wherein the stress response is more manageable. This is accomplished by a change in geometry, sometimes small, other times large.

I offer an example that illustrates the point. Consider a child's balloon as a structure, which, when inflated, is as flexible a structure as one is likely to see. Imagine squeezing this flexible latex membrane structure between two fingers and a mental image of an altered shape results. It is now fair to inquire as to the nature of the resisting forces within the membrane and the subsequent internal stress response. The answer to such inquiry is that only stresses parallel to the surfaces of the membrane result; in the case of this example, tension only. In fact, in classical shell theory, these stresses are called membrane stresses. In our work, with the load applied from without a flexible pipe, rather than from within (as in the case of the inflated balloon), compression rather than tension results. In the pipe business we call this membrane stress response, ring compression.

One may further inquire as to the bending response. After all, in the case of the flexible pipe, does not the deformation induced by earth and other service loads create bending in the pipe wall? Since bending excites both tensile and compressive stresses, how does one account for the tensile stresses? Well, in a true membrane, bending stress is never a response. And although a flexible pipe is more stiff than a true membrane, the ability to sufficiently deform, in compliance with the applied loads, insures a resulting geometry of structure such that favored membrane stresses of compression dominate the response. The less favored bending stresses of tension and compression are of lesser significance.

Let's get a feeling of size of the numbers we call stiffness. The classical definition is as follows: *STIFFNESS IS THE FORCE REQUIRED TO EXCITE A UNIT OF DISPLACEMENT.* This measure is achievable either by the requirement for plastic pipes, the parallel plate test ASTM D2412, or by calculation, the requirement for metal pipes. The calculations simulate the loading, the geometric characteristics and the material properties. In either case, that which is sought is:

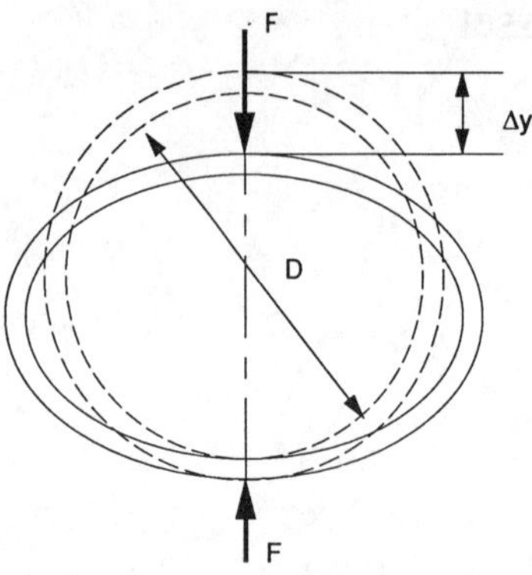

Figure #1

Pipe Stiffness = PS = F/Δy

where F is the measure of force applied on the outer surface at each end of the vertical diameter connecting crown and invert, and Δy is the measure of the total shortening of that same diameter, when Δy = 5% of the inner diameter of the pipe. See Figure 1.

When calculated, rather than being determined by test:

Pipe Stiffness = PS = $E*I/(0.149*r^3)$

which is an outcome of classical elastic analysis. In this case, note that there are two elements of stiffness; a material stiffness, E, and a geometric stiffness I/r^3. The pipe stiffness is proportional to the increase of the moment of inertia, I, or to the decrease of the cube of the radius, r^3.

On the flexible end of the scale, there are pipes that are manufactured with a pipe stiffness as low as PS = 10 lb/in/in. On the rigid end of the scale, if the above analytical expression is applied to calculate the pipe stiffness of a standard reinforced concrete pipe (RCP) of an arbitrarily selected 48 inch nominal diameter, the result is a pipe stiffness in the neighborhood of PS = 6000 lb/in/in. Note the greater compliancy of the flexible pipe. The large measure of resistance to deformation developed by the rigid pipe insures that the less favored bending mode of structural response will dominate, thereby inducing large bending stresses which includes a high threshold of tension requiring ductile steel reinforcing to assist the brittle concrete to perform as a

competent structure. The limit of performance is a function of the onset of cracking as determined by the standard Three Edge Bearing test. Of course, in application, neither the flexible pipe nor the rigid pipe should be expected to be loaded only with concentrated loads at opposite ends of a diameter.

The word compliance, which I have used heretofore in a general conversational manner, has a formal definition in that field of study known as the mechanics of solids. Compliance is the inverse of stiffness. Keeping the sense of this definition, I define pipe compliance as follows: *PIPE COMPLIANCE IS THE INVERSE OF PIPE STIFFNESS*. The noted flexible pipe has a pipe compliance in the neighborhood of 100×10^{-3} and the noted rigid pipe has a compliance in the neighborhood of 0.2×10^{-3}. A highly compliant pipe will develop less bending stress and thereby require less section modulus to manage the bending stress. A lower section modulus implies a lower moment of inertia – a requirement of consistent with the concept of a flexible pipe. So it turns out that a thin pipe of low geometric stiffness furnishes, and thereby requires less of a section to handle the resulting bending moment. The more compliant the pipe, the more efficiently the bending moments are eliminated. And that is only the beginning.

I now call your attention to the geometry of the pipe cross-section and its effect on the moment response. Every designer of thin structural shells, of which a pipe is just one of many such shells, understands that a sharp change in the curvature of the shell gives rise to sharp bending moments, in the vicinity of the change – and that is precisely that which should be minimized.

To illustrate, imagine a closed cylindrical pressure vessel capped at both ends with a hemisphere of the same wall thickness as that of the cylinder. If a longitudinal cross-section is cut, the walls of the cylinder, indicated by top and bottom horizontal lines, are tangent to each end of each semicircle representing the walls of the hemispherical end caps. See Figure 2.

At each point of tangency, the infinite radius of the straight lines of the walls of the cylinder ρc, meets and changes abruptly to the small (by comparison), finite radius of the wall ρs, of the hemisphere. The curvatures of each, by definition, the inverses of the radii, experience a comparable, but inverse, relationship; large curvature in the case of the hemisphere wall and small curvature (actually = 0) in the case of the cylinder wall. These curvatures meet at the tangent intersection of the two walls. Imagine

2

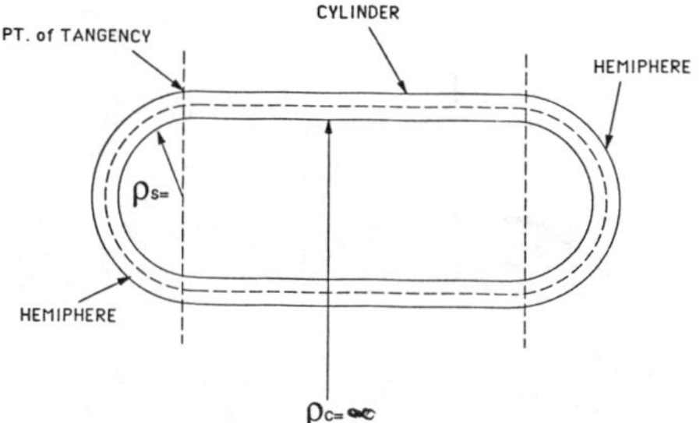

Figure 2

now that the pressure vessel is now charged and it expands – but does not do so uniformly. At the intersection of the hemisphere and the cylinder, each geometric shape wants to expand in a manner consistent with its own geometry. The cylinder wants to swell and become a larger cylinder; the hemisphere wants to open as petals of a flower. Being fastened, one to the other in a manner designed to maintain the tangency of their intersection, they inhibit each other's expansion. This incompatibility gives rise to very significant shears and bending moments when compatibility of the deformations is enforced in the walls of both cylinders and hemisphere in the vicinity of the joint.

If the principle of the relationship between curvature and moment is applied to that of the pipe or culvert, then one concludes that, other things being equal, a cross-section of constant curvature; that is, a circular cross-section, is the most efficient geometry from the structural point of view of stress response in the plane of the cross-section. This is good news for all manufacturers of circular pipes.

We now turn to a study of the loads on a buried pipe or culvert. It is first important to establish the fact that it is not only the magnitude of the load, but also the character, or distribution, of the load acting on the pipe that determines the stress and deformation response of the structural pipe. We begin with the magnitude of the load.

Imagine, if you will, four piano movers carrying a grand piano down a spiral staircase. Think of each as a point of support for the piano at any moment in time; think of the piano as a plane structure. Since only three points of support are necessary for the geometric stability of a plane, it may well be the case that one of them is faking his or her, share of the burden. This latter person, the one only pretending to carry the load, may have studied engineering at any one of our nation's universities and excelled in his or her courses in structures. If so, he or she was able to conclude that four supports results in one redundant support, and there is an opportunity to benefit from the knowledge that the three others, if strong enough, will offer sufficient reaction forces to maintain the stability of the piano. The pretender offers the least support, that is, the least stiffness for the support of the load. The piano is very smart, however, and when it senses the relaxation of the fourth support, it leans more heavily on the three stiffer supports. No matter what the appearance of the faker, bent over and deformed as he may be, he carries little load, little shear, and little moment. And the reward is less vulnerability to pulled muscles, torn ligaments and herniated disks.

By analogy, I use this example to explain the attribute of the flexible pipe in a highly redundant soil mass. Working as a soil-structure composite both the pipe and the soil mass will support a superimposed dead load of embankment and traffic live load. Should the pipe have a stiffness less than the surrounding soil, a likely event with corrugated metal and plastic pipes, then the load will direct itself to the stiffer soil where it will look for support. This is the mechanism for the development of the soil arch. On the other hand, should the pipe be stiffer than the surrounding soil, that is, it is less compliant than the surrounding soil, more than a fair share of load will be attracted to the pipe. In fact, this rigid inclusion, which may be a reinforced concrete pipe, had better be very strong, with very thick wall and lots of steel reinforcing, for it may have lots of additional load to take. Of course, the very act of providing all this required additional strength further stiffens the pipe which, in turn, attracts even more load. I hope you will agree that flexible pipes offer great advantage in creating those conditions for a low level of load to be attracted to the pipe. The more compliant the pipe, the less the load attracted to the pipe. My suggestion to all of us here is that, in small increments, we work to drive down the manufactured stiffness of the pipe flexible and rigid alike, and simultaneously work to drive up the stiffness of the compacted soil envelope.

Now let's consider the character of the load, that is, the nature, or geometry, of the distribution of the load around the circumference of the pipe. At the soil pipe interface, the forces that exist, whatever their orientations, may, for purposes of analysis, be described in terms of a normal force component and a tangential force component; also known as surface tractions. The following comments are directed to the forces of surface tractions that exist at the pipe–soil interface.

Consider the circular ring of the cross–section of the buried pipe, itself a cylinder of long length. If the surface tractions were all radial and of the same magnitude (ideal and realistically unattainable) then there would be no bending in the pipe wall, only the favored ring compression; no shear, no moment. This is true no matter what the material, profile or thickness of the pipe wall. This is why submarine hulls can be thin shells when the cross–section is largely a circle.

Picking up on this principle, it is true that the more non–uniform the distribution of load, the greater the reacting shear and moment forces within the pipe walls. The very worst possible loading of stable equilibrium is when there are only two equal and opposite concentrated loads on opposite ends of a diameter of the pipe when viewed in cross–section. Isn't it interesting that this loading, which causes the greatest measure of bending stress, is the basis for the parallel plate test for stiffness rating of plastic pipes, the analysis for stiffness rating for flexible metal pipes, and the three–edge bearing test for the strength ratings of rigid reinforced concrete pipes. Assuming static equilibrium satisfied, any other load distribution of the same resultant force would serve to reduce the bending moments in the pipe wall.

It is for this reason that, for all classes of pipe, good uniform compaction at all locations around the pipe is a desirable condition. This will promote gradual point–to–point change in the character of the load which, in turn, precludes the development of high moments which will accompany sharp changes the geometry of the loading. It is interesting to note that reducing the stiffness of trench backfill, by introducing highly compliant material in the vicinity above the crown of the pipe, is a strategy designed to reduce the load coming to the crown of rigid concrete pipe (which would otherwise attract large loads due to the stiffness of the pipe). It is not necessarily the case, however, that this results in a reduction of the pipe wall moment, precisely because of the abrupt change in the loading geometry due to the abrupt change in the soil stiffness.

Taking the concept of good compaction a step further, good firm support at springline pays high dividends in that the horizontal forces of reaction serve to reduce the bending moment in the wall from what it would be in the absence of this support. The benefits of this effect are explained, in part, by analogy with the curved beam.

Imagine that the top half of a pipe is a curved

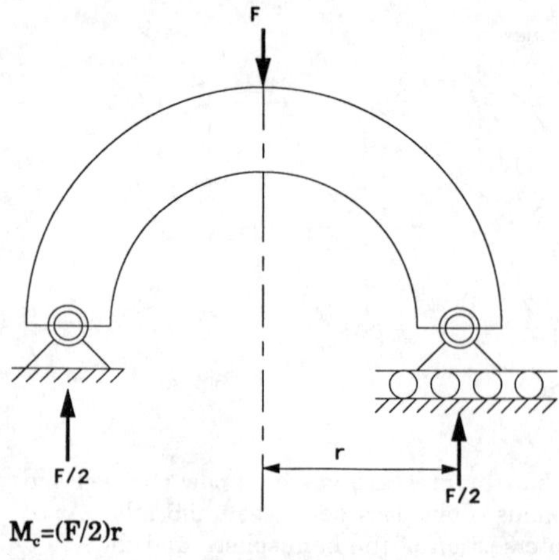

$M_c=(F/2)r$

Figure 3   Curved Beam

circular beam, supported on the left by a pin support and on the right with a skateboard to simulate the absence of lateral support at springline. See Figure 3. A single concentrated load at the crown will be sufficient loading to establish the principle that, with the application of the load, the right support moves out freely and only the vertical reaction at each support survives. The moment at the crown is one vertical reaction times the horizontal distance to that point (the radius of the pipe).

If, however, horizontal reactions on the pipe were permitted to develop due to competent support at springline, the analogous curved beam now an arch would have the benefit of this horizontal reaction creating a moment of opposite sense (to that created by the vertical reaction) about the crown. See Figure 4. This is the benefit to be captured by good compaction at spring line. (The analogy is incomplete in that there are also moments in the pipe wall that should be introduced at the curved beam reactions. These moments also serve to reduce the moment at the crown and their exclusion from the analogy does not compromise the illustration of the principle.)

Consider a flexible pipe bedded in a trench and then properly backfilled with appropriate compaction. With its capacity to deform into an oval shape, the lengthing of its horizontal diameter will forcibly additionally compact the soil at springline and thereby improve the passive pressure in that vicinity. The result is a reduction of moment in the pipe wall to the point where ring compression stresses could dominate the response. A similarly bedded and backfilled rigid reinforced concrete pipe due to its inability to sufficiently change shape, will

4

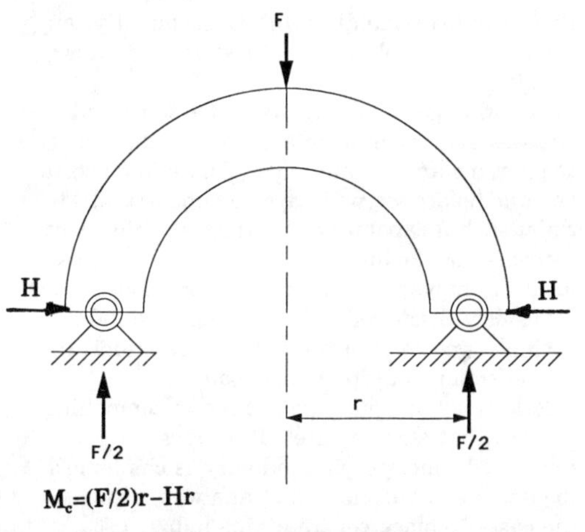

$$M_c = (F/2)r - Hr$$

Figure 4   Arch

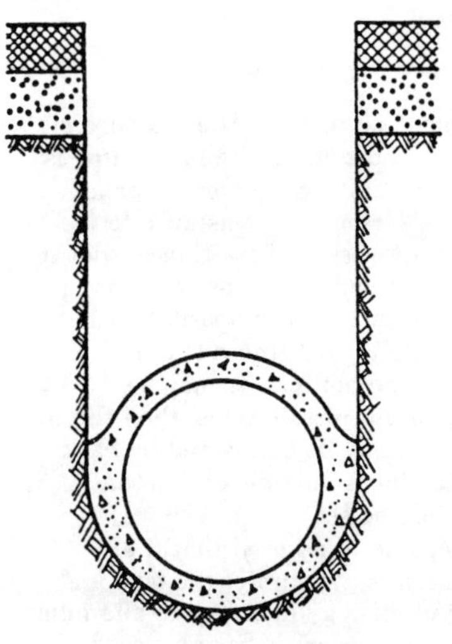

Figure 5

experience no such gain; it depends upon the inherent moment capacity of the wall section for its performance. Bending stresses dominate the response.

To further illustrate the advantage of the development of significant passive pressure at springline, I call your attention to cast-in-place concrete pipe which is well known in California, Arizona, and some parts of Nevada, New Mexico and Texas. See Figure 5

This pipe is produced by a continuous cast process against the stand-up vertical walls and bottom of a trench. The pipe bears directly against the side walls of the trench, which in turn immediately activates, and insures the development of passive pressure in response to load, even at small enlargements of the horizontal diameter. That which relates to our discussion here is that this pipe is of unreinforced concrete with sizes up to 120" nominal diameter and with wall thicknesses just a bit thicker than reinforced concrete pipe. The concrete is generally specified at 3000 psi. The point being made is, that *with properly developed passive pressure at springline, assured by intimate contact with the undisturbed trench walls, this massive rigid pipe responds in the ring compression mode.* If bending were to dominate, which it doesn't, surely reinforcement would be required to handle the flexural tension. Reinforcement is not used with this pipe. If one questions the veracity of the argument, it is noted that approximately 11,000,000 feet of this pipe, with 22% of 54" nominal diameter, and larger, has been constructed since 1955. The conclusion is that all pipe, flexible and rigid, benefit greatly from competent and adequate lateral support at springline.

Getting back to deflections, one of the problems that has plagued the profession has been the lack of confidence in the ability to predict deflections in advance. The ring deflection theory, which works well for some applications, works poorly for others. Noble efforts have been made to bring contradictions into alignment by altering values of E', the presumed soil stiffness, and by anticipating construction quality with the addition of inspection factors. And yet contradictions exist. The culvert design profession has all but abandoned this formula; the plastic pipe industry is currently doing the same.

Because it would be of great value to be able to predict deflections in advance, with a sense of surety, it may be useful to explore the possibility that the basic relationship, the Modified Iowa Formula, includes an inappropriate assumption. Let's start from the beginning.

Using the definition of the term 'stiffness', the relationship between force, stiffness and displacement is:

$$\text{displacement} = \frac{\text{force}}{\text{stiffness}}$$

This is interpreted by the Modified Iowa Formula to mean:

$$\text{displacement} = \frac{\text{load on the pipe}}{\substack{\text{pipe} \\ \text{stiffness}} + \substack{\text{soil} \\ \text{stiffness}}}$$

or,
$$\Delta = \frac{F}{K_p + K_s}$$

The displacement is taken as the measure of vertical diameter shortening; the numerator is measure of a force, and; the denominator is stiffness, the measure of the resistance to deformation. It appears, at first glance, that it is very reasonable to add the pipe stiffness to the soil stiffness to arrive a composite stiffness. Furthermore, if the pipe stiffness is presumed to be calculable and the deflections may be measured for existing installations, then the in-place soil stiffness may be discovered by back calculation. Standards for soils of different qualities may then be set.

But is it reasonable to assume that the composite stiffness is the sum of the pipe and soil stiffness? An analogy is drawn to a familiar formula, Ohm's Law, wherein for an electrical resistance circuit:

$$E = IR, \text{ or: } I = \frac{E}{R}$$

where E is itself a force and R is the resistance offered to the passage of the current I. *The stiffness (K), the resistance to deformation in the pipe deflection equation, is the analog of the resistance (R) to the passage of current in the electrical equation.* The coupling of two stiffness in the denominator of the pipe deflection equation is assumed to be additive. But how would one couple two resistances, say R_1 and R_2, in an electrical circuit? We now have two choices: additive, in series, as in the pipe deflection equation; or in parallel, an opportunity not offered in the pipe deflection equation.

In an earlier work, reported on at the ASCE Conference on Underground Plastic Pipe (New Orleans, 1981), Mr. Howard Blower and I were able to demonstrate that the coupling of the stiffness of the pipe, the sidefills adjacent to the pipe, the backfill above the pipe and the soil below the pipe, present a coupling arrangement, not at all describable by the simple additive denominator of the pipe deflection equation. Note, however, that if the stiffness of the soil envelope dominates the denominator of the pipe deflection equation, as is the case for flexible pipe of low stiffness (relative to the soil), then any error in the coupling of the pipe and soil stiffness is of lesser, or little, consequence. This was the case for the early studies on flexible pipe by Marston and Spangler. The difficulty in predicting the deflection of semi-rigid pipe, using the Modified Iowa Formula, is due, in large part, to the fact that the stiffness of such pipe is not necessarily low when compared to that of the surrounding soil. Consequently, any error in the coupling of stiffness is of greater significance.

A knowledge of anticipated deflections and stresses are essential to a proper prediction of the performance of any pipe. There is much to be done before we will have the opportunity to claim such a capability. Fortunately, the stress response for flexible pipe quite properly lends itself to analysis by the ring compression theory, previously discussed. The flexible pipe industry is challenged to offer a consistent relationship for deflections of pipe under load.

Deflections of rigid pipe are not of compelling interest, but stresses are! The precast reinforced concrete pipe industry is challenged to offer a general method of analysis, as does the cast-in-place concrete pipe industry, wherein a stress response is predicted for anticipated service conditions.

The manufacturers of semi-rigid pipe are challenged to offer a general method of analysis wherein both stress and deflection responses are a reflection of both the ring compression and bending moment in the pipe wall.

Engineers in the plastic pipe industry are challenged to learn more about the advantages of increasing pipe flexibility wherein the attribute of the time dependent property of stress relaxation enables the transfer of load with time, to the soil part of the soil-structure composite. A knowledge of load and stiffness at any point in time should be favored over the current strategy of design based upon early loads and later stiffness.

Public officials and manufacturers, alike, are challenged to develop quality assurance programs for field installation procedures that would enhance the performance of all pipes. This would enable the earlier introduction and acceptance of new products.

Engineers involved in public and private practice are challenged to become familiar with the more general use of computer analyses, such as CANDE, wherein both stresses and deflections may be calculated for a great variety of conditions of pipe, surrounding soil, loads, and conditions of embedment.

And finally, academics are challenged to persuade their colleagues, and institutions, that serious study in soil-structure interaction should be a part of basic study in every Civil Engineering program,

Structural Performance of Flexible Pipes, Sargand, Mitchell & Hurd (eds) © 1990 Balkema, Rotterdam. ISBN 90 6191 165 6

Rib-stiffened soil-steel structures

George Abdel-Sayed
University of Windsor, Windsor, Ont., Canada

Shantaram Ekhande
Flint Automotive Division, G.M., Great Lakes Technology Center, Flint, Mich., USA

ABSTRACT: A simplified procedure is presented herein for the analysis and design of circumferentially stiffened soil–steel structures. The procedure is based on the results of three–dimensional analysis of stiffened conduits using the method of finite strip. The conduit is divided into stiffened and non–stiffened zones in which the width of the stiffened zone is taken to be equal to that of the stiffener.

During construction, the lateral soil pressure is found to be carried by the stiffened and non–stiffened zones in proportion to their rigidity. On the other hand the thrust, bending moment and deflection, induced after backfill is completed, by superimposed dead and live loads can be calculated in two steps. Using plane strain analysis, their magnitude is calculated and then modified by factors γ_n, γ_m and γ_d, respectively. These factors are evaluated using the 3–D finite strip analysis, i.e. they account for the interaction between the stiffened and non–stiffened zones.

1 INTRODUCTION

Composite soil–steel structures are economical alternative to the classical short–span bridges. They are built of flexible corrugated steel shell embedded in soil. With the increase in span, problems arise during construction due to the large deformation of the conduit, as well as under dead and live loading because of the low buckling strength of the shell.
The building of long span soil–steel structures is made possible and their performance has been improved by providing circumferential stiffeners, Fig. 1.

However, the present codes and design practices do not provide any procedure to evaluate the effect of these stiffeners (Ontario Highway Bridge Design Code OHBDC, 1983, American Association of State Highway and Transportation Officials, AASHTO Bridge Specifications, 1983). All the methods presently available for the analsis and design of composite soil–steel structures are based on considering a slice of a unit width of the conduit and the surrounding soil in which the shell action of the structure is neglected. This approach is valid for non–stiffened conduits, with constant loading in the longitudinal direction. Also, the OHBDC, 1983 provides a formula for the maximum allowable rise at the crown during construction. This formula provides a limit on vertical deflection so that the stresses in the conduit wall remains below its yield limit. Conduits with stiffeners are not covered by this formula.

Therefore, procedures developed using three–dimensional analysis are presented in this paper for the evaluation of the composite action of the stiffeners and the conduit panels. The stiffeners effect is outlined so that the simplified method of plane analysis can be used together with modifiying factors to calculate the internal force components in stiffened conduit walls.

2 THE EFFECTIVE WIDTH

The stiffeners and shell are bolted together and assumed to be monolithic. The rigidity of the composite section can be evaluated by considering an imaginary width of the shell to be effective with the stiffener. Two different formulas were developed (Ekhande and Abdel–Sayed, 1986) to calculate the effective width. One formula is based on equivalent rigidity and the other on equivalent stresses. That is to say that maximum rigidity or maximum

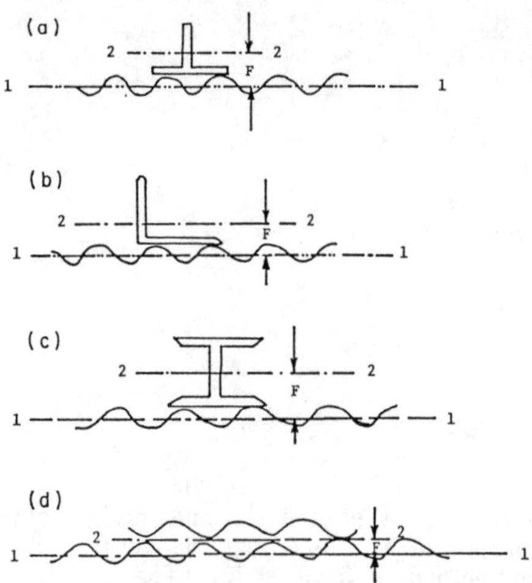

Fig. 1 Circumferential Stiffeners: (a) Tee;
(a) Angle; (c) I-beam; and (d) corrugated
sheets. 1-1 reference surface of the shell;
2-2 centroidal surface of the stiffened element

stresses calculated by considering the stiffener with the effective width should be of the same magnitude as those obtained from proper three-dimensional shell analysis.

It has been found (Abdel-Sayed and Ekhande, 1985) that the effective width is relatively small when compared with the spacing between the stiffeners or with the effective width of stiffened plane panels. This fact falls in line with the characteristics of curved beams in which the curvature of the flange leads to reduction in their effective areas and the total bending rigidity of curved beams (Seely and Smith, 1967). These results point also at the poor shell effect which is attributed to the low axial rigidity of the corrugated shell in the longitudinal direction.

Therefore, it has been concluded that a reasonable approximation can be made by considering the effective width to be equal to the width of the stiffener.

3 ANALYSIS

The method of finite strip is selected for the analysis since it demonstrates great computing economy when compared with the method of finite element (Ekhande, 1986). The strips are oriented in the curved direction so that the effect of stiffeners is determined by considering the stiffeners width as individual strips. The stiffened and non-stiffened strips are treated

as if being made of orthotropic material in which each strip properties are calculated as the average mechanical properties of its section (Abdel-Sayed, 1971). It is to be noted that an eccentricity 'F' exists between the centroidal surfaces of the stiffened and the non-stiffened zones of the shell. Therefore, the displacement functions chosen for the stiffened strips incorporate the eccentricity F and the displacement degrees of freedom along the centroidal surfaces of the main shell. This approach satisfies the continuity between the main shell strip (non-stiffened zone) and stiffened strip. (Ekhande, 1986)

The surrounding soil is represented by radial and tangential extensional spring elements acting over the shell. The spring coefficients are calculated by using the coefficients of soil reactions as proposed by Okeagu and Abdel-Sayed, (1984). These coefficients are dependent on the type and depth of soil, as well as on the direction of displacement of the shell wall.

The three-dimensional analysis is directed to examine and develop a rational procedure to determine the effect of stiffeners on the behaviour of soil-steel structure. It addresses these effects at two main stages in the life of the structure, namely a) during construction and b) under dead and traffic loads.

(a) Stiffened Conduits Under Construction Loads

During side filling the upper zone of a conduit has no soil cover and moves upward. Its behaviour can be simulated as a cylindrical shell subjected to horizontal forces lateral to its edges Fig. 2a. The effect of the area of stiffeners area and their spacing is found to be as follows:

a-1 Stress Distribution

Figure 3 shows the maximum stresses induced in the bottom extreme fibre of non-stiffened and stiffened (with corrugated sheet or with Tee-section stiffeners at S = 686 mm, with $\alpha = 83.75^{\circ}$, R = 873 mm) subjected to the same magnitude of horizontal loads. The stress reduction in non-stiffened zone is very significant. It is reduced from 3200 psi to 1500 psi and to 1000 psi due to corrugated and tee type stiffeners respectively, which constitute a decrease in stresses of about 53% and 68% respectfully. However, less reduction is observed in the stresses of the stiffened zone. Furthermore, the corrugated sheet stiffeners which accounted for 19% increase in the shell material, reduced the maximum stress by 20% in the stiffened zone, while the Tee–section stiffeners (W = 101 mm) which account for 35% increase in the shell material, reduced the maximum bottom fibre stresses by only 21%. This indicates that for a given spacing of stiffeners the stiffener's size does not have significant effect on reducing the maximum stresses.

The effect of spacing on the stresses and deflections in the upper zone of a full scale structure with and without thrust beams is shown in Table 1, under the action of identical horizontal loads, while the totals cross sectional area of the stiffeners is the same in all cases. The columns (a) and (b) of Table No. 1 show that the state of stress and the deflection at the crown level is the same for structures, with and without thrust beams. The reductions in combined stresses under the stiffeners at spacings of 2.75 m (9'), 1.83 m (6') and 1.0 m (3') c/c are 44%, 55% and 65% respecitvely. This indicates that the spacing of the stiffeners has more pronounced effect on the stress reduction than the type of stiffeners and their total area of cross–section.

a–2 Crown Deflection

Table 2 shows the crown deflection for stiffened and non–stiffened shell under same magnitude of horizontal load. In a Test conduit with corrugated stiffeners the crown deflection is reduced by 55% while the same in a shell with Tee–section stiffeners is reduced by 73%. Also, more reduction in deflection is achieved (Table 1) when reducing the spacing while keeping the total area of siffeners unchanged.

It is concluded that, in general, the stiffeners are more useful in controlling the crown deflection during construction when compared to the reduction in the stresses. However, if the stiffeners are closely spaced, both deflection control and bending moment capacity are improved.

a–3 Proposed Procedure for Analysis and Design

Table 2 shows the crown deflection to be almost equal for the stiffened and non–stiffened zones. This means that the

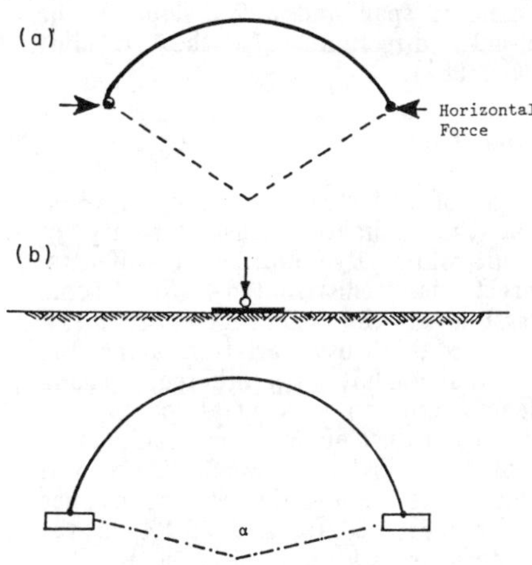

Fig. 2   Models to Simulate Soil-Steel
Structures: (a) During backfilling;
(b) Under traffic live load

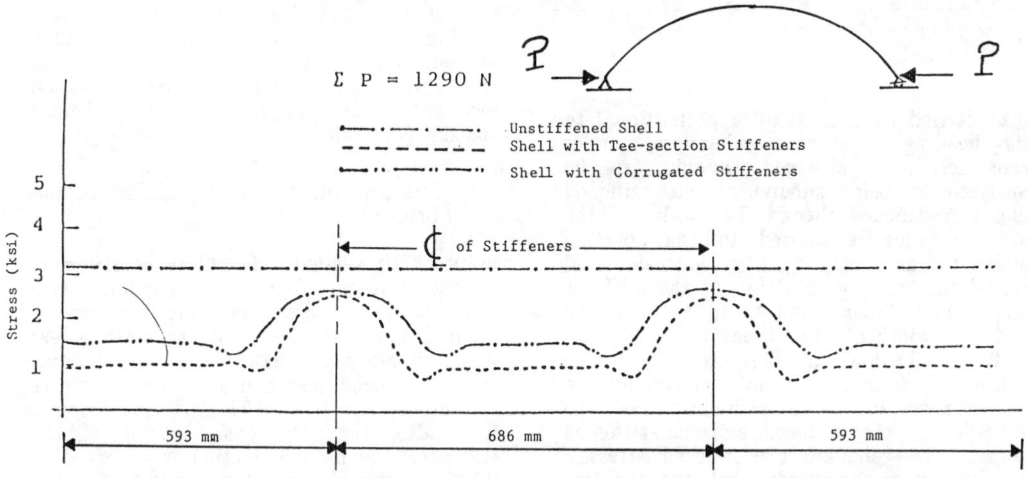

Fig. 3. Bottom fibre stress in stiffened and non-stiffened model under horizontal loading P (Ekhande 1986)

Table 1. Effect of stiffener spacing on the structure behaviour during side filling

| Variable | | Structure Type | | | | | |
| | | No stiff. | S = 2.75 m | | S = 1.83 m | | S = 1.0 m | |
		a & b	a	b	a	b	a	b
Stiffener Material Added (%)		-	72	72	70	70	70	70
Combined Stress at (MPa)	Unstiff. Zone	7.4	1.31	1.31	1.17	1.23	1.01	0.97
	% Drop	-	82	82	84	83	86	87
	Stiff. Zone	7.4	4.11	4.12	3.26	3.31	2.59	2.59
	% Drop	-	44	44	56	55	65	65
Crown Deflection (mm)	δ	7.5	1.32	1.27	1.24	1.22	1.12	1.04
	% Drop	-	82	83	83	84	85	86

R = 4700 mm, α = 80°, Σ P = 1121 N

a = structure without thrust beams

b = structure with thrust beams

Table 2. Deflection comparison for stiffened and non-stiffened shell during side filling

| Variable | | Stiffener Type | | |
		No stiff.	Corrugated	Tee-section
Stiffener Material Added (%)		-	19	35
Total Rigidity		0.098 in^4	0.2145 in^4	0.7215 in^4
Deflection at (mm)	Unstiff. Zone	3.65	1.61	1.04
	% Drop	-	55	71
	Stiff. Zone	3.65	1.57	0.97
	% Drop	-	55	73

R = 873 mm, α = 83.75° Σ P = 1290 N, S = 686 mm c/c

load carried by each strip is proportional to its bending rigidity. Therefore, under construction, a stiffened conduit can be analysed by being subdivided into stiffened and non-stiffened slices. The width of the stiffened zone is limited to that of the stiffener, and its rigidity is calculated considering composite action of the stiffener and the shell connected to it, i.e. effective width = width of the stiffener.

The deflection at the crown can be calculated considering the rigidity of the upper zone to be a summation of the rigidities of the stiffened and non-stiffened zones. To calculate the induced stresses, the load is to be divided between the two zones in proportion to their rigidities.

(b) <u>Stiffened Conduits Under Gravity Loads</u>

The circumferential stiffeners provide variable rigidity to the soil–steel structure along its longitudinal direction. It is anticipated that the soil loading would vary over the length of the conduit. This variation could be more pronounced under deep cover due to the cross–arching of the soil between stiffened zones. However, the present paper is based on analysis considering uniform load distribution above the conduits. The live load is considered to be dispersed in the soil with a 1:1 slope in the conduit span and a 2:1 slope in the longitudinal direction of the conduit (OHBDC, 1983).

b–1 Axial Thrust

The design of soil–steel structures is often based on the ultimate thrust capacity of the conduit wall. By addition of stiffeners, the thrust is redistributed with some concentration at the stiffeners. Figure 4 shows the axial thrust variation along the shoulder of a conduit with different spacing of stiffeners due to a central patch live load of two axles of 32 kips each at a spacing of 4 ft., Ekhande, 1986. The axial thrust at the non–stiffened zone has dropped by 6.8%, 20.9% and 29.1% in the case of stiffener spacings of 2.75 m (9'), 1.83 m (6') and 1.0 m (3') c/c respectively (Table 3). This indicates that the closer the spacing of stiffeners, the more reduction is achieved in axial thrust in the non–stiffened zone along the haunch. On the other hand the axial thrust increases at the crown with the decrease in stiffner spacing (Fig. 5a). In reference to bending moment and deflection, it is noticed that the magnitude of bending is reduced at all locations with the decrease of stiffeners spacing (Fig. 5b)

b–2 Stresses Due to Combined Thrust and Bending

In order to evaluate the effect of stiffeners on the extreme fibre stress, conduits with and without stiffeners are examined. Figures 6 and 7 show the combined stress variation along the haunch and crown level of the stiffened and non–stiffened structures respectively under two 32 kips axle loading. Here also, the closer the spacing of the stiffeners, the more reduction is observed in the combined stresses at non–stiffened zone.

b–3 Modification Factor

A large number of soil–steel structures were analysed, in order to establish the structural effects of the various parameters; S = the spacing between stiffeners; F = the eccentricity of the stiffened zone; W = the width of the stiffened zone; α = the intersection angle of the conduit segment under consideration; EI_1 = the bending rigidity per unit width of the non–stiffened zone; and EI_2 = the bending rigidity per unit width of the stiffened zone. The maximum values of axial thrust, N, bending moment, M, and deflection, δ_c, were calculated considering non–stiffened structure. Then individual unit strips of stiffened and non–stiffened zones were analyzed under the same loading. The maximum values of thrust, bending moment and deflections are listed in Table 4. Modification factors (γ_n, γ_m and γ_d) were calculated for N, M and δ_c, respectively at the stiffened and non–stiffened zones, as follows: factor, γ =

$$\frac{\text{3–D Actual value of entity (e.g. } N, M, \delta_c)}{\text{Result of non–stiffened strip analysis}}$$

$$(1)$$

A linear regression curve fitting technique was then used to obtain the factors γ_n, γ_m, and γ_d in terms of the following non–dimensional parameter: (S/R), (F/f), $\frac{1 - \cos \alpha/2}{2 \sin \alpha_2}$ $(= \frac{\text{rise}}{\text{span}})$, and the stiffeness

ratio $(\frac{W \cdot EI_2}{W \cdot EI_2 + (S-W) \cdot EI_1})$.

The modification factors have the following form:

$$\gamma = k(\frac{S}{R})^a (\frac{F}{f})^b (\frac{1-\cos \alpha/2}{2 \sin \alpha/2})^c$$

$$(\frac{W \cdot EI_2}{W \cdot EI_2 + (S-W) \cdot EI_1})^d \qquad (2)$$

Where k, a, b, c, and d are constants tabulated in Table 5.

b–4 Procedure of Analysis and Design

A plane analysis (using unit width) of stiffened and non–stiffened zones is

Table 3. Effect of Stiffener Spacing on the Soil-Steel Structure Under Patch Live Loading

Parameter / Ratio of Rigidity		No. Stiff. S = ∞	Structure Type S = 2.75 m	S = 1.83 m	S = 1.0 m
Combined Stress Haunch	MPa	−43.03	−40.48	−34.96	−31.66
	% Drop	–	5.9	18.7	26.40
Axial Thrust Haunch	N/mm	97.9	91.24	77.40	69.35
	% Drop	–	6.8	20.9	29.1
Bending Moment Crown	N.m/mm	1.014	0.965	0.863	0.792
	% Drop	–	4.8	14.9	21.9
Bending Moment Haunch	N.m/mm	−0.721	−0.703	−0.623	−0.569
	% Drop	–	2.5	13.6	21.1
Crown	mm	3.5	3.35	13.7	20.3
Deflect	% Drop	–	4.3	13.7	20.3
Average % Drop		–	4.9	16.4	23.8

$R = 4700$ mm, $\alpha = 80°$, $H_c = 0.91$ m, $q = 19.15$ kN/m^2

performed under actual loading; assuming that the stiffened and non–stiffened zones are not interacting with each other. Then the modification factors 'γ' are calculated for N, M and δc for both stiffened and non–stiffeened zones using Eq. 2 and Table 5. The effect of stiffened and non–stiffened zone interaction is accounted for by multiplying the results of plane analysis by the corresponding factors. Table 5 also shows the variations of error which is low within the non–stiffened zone but relatively high for the stiffened zone.

CONCLUSION

A simplified procedure is outlined for rib–stiffened conduits in order to determine the magnitude of the maximum thrust, the stresses and/or deflection. It is based on linear three–dimensional analysis of stiffened conduits under construction loadings and under gravity dead and live loads.

Under construction, the effects of horizontal soil pressure can be analysed as being distributed between the stiffened and non–stiffened strips in proportion to their bending rigidities.

Under gravity dead and live loading, a plane analysis can be applied and its results

Table 4. Modification factors γ_n, γ_m, γ_d calculated from finite strip analysis

Condition				Parameter (in., degress, ksi)				Non-Stiffened Zone			Stiffened Zone		
NO	R	α	S	F/f	W	EI_1	EI_2	γ_n	γ_m	γ_d	γ_n	γ_m	γ_d
1	185	80	108	2.182/1	10	2.089E6	9.956E7	0.946	0.937	0.94	4.58	2.93	2.88
2	185	90	108	2.66/1	10	2.089E6	10.10E7	0.93	0.92	0.922	4.60	2.48	2.48
3	185	80	72	1.986/1	10	2.089E6	7.62E7	0.873	0.824	0.834	3.77	2.56	2.51
4	200	90	72	2.503/1.	10	2.089E6	7.856E7	0.885	0.860	0.868	3.57	2.01	2.02
5	453	80	72	5.037/1.07	10	4.783E6	16.12E7	0.86	0.895	0.86	3.37	1.17	1.25
6	400	90	72	4.42/1.07	10	3.414E6	12.62E7	0.838	0.89	0.90	3.76	1.33	1.39
7	350	100	72	3.504/1.07	10	2.721E6	5.869E7	0.81	0.90	0.91	3.52	1.31	1.37
8	300	120	90	1.0/1.0	30	4.178E6	3.054E7	0.956	0.96	0.958	1.34	1.09	1.09
9	250	100	90	1.25/1.25	30	4.66E6	3.155E7	0.93	0.93	0.937	1.41	1.15	1.15
10	185	80	36	1.372/1.0	4	2.089E6	2.858E7	0.90	0.893	0.90	3.14	1.95	1.98
11	160	85	36	1.6/1.0	4	2.377E6	3.177E7	0.88	0.875	0.885	2.97	1.97	1.98
12	185	80	36	2.166/1.0	6	2.089E6	5.702E7	0.86	0.78	0.797	3.26	2.137	2.142
13	210	90	36	2.50/1.00	6	2.377E6	6.164E7	0.92	0.87	0.88	3.47	1.57	1.60
14	210	100	108	3.34/1.2	10	4.294E6	11.35E7	0.908	0.90	0.90	4.24	1.94	1.93
15	275	95	36	4.27/1.25	4	6.528E6	25.44E7	0.99	0.727	0.73	9.97	2.18	2.21
16	195	80	67	2.06/1.0	5	2.089E6	9.48E7	0.887	0.867	0.876	4.59	3.23	3.18
17	170	100	67	2.88/1.0	5	2.089E6	9.934E7	0.885	0.885	0.89	4.49	2.538	2.55
18	180	80	67	3.295/1.1	5	2.876E6	10.69E7	0.876	0.857	0.865	4.32	2.77	2.73
19	210	90	67	3.598/1.2	5	4.214E6	12.39E7	0.88	0.86	0.868	5.29	2.14	2.14
20	240	100	67	4.058/1.3	5	7.061E6	18.16E7	0.867	0.833	0.84	6.798	2.14	2.11
21	165	80	108	2.182/1.	10	2.089E6	9.956E7	0.96	0.956	0.96	4.43	3.06	2.96
22	215	80	108	2.182/1.	10	2.089E6	9.95E7	0.926	0.914	0.92	4.74	2.70	2.68
23	165	80	72	1.986/1.	10	2.089E6	7.62E7	0.866	0.82	0.84	3.67	2.73	2.64
24	215	80	72	1.986/1.	10	2.089E6	7.62E7	0.88	0.84	0.85	3.67	2.30	2.29
25	165	80	36	1.372/1.	4	2.089E6	2.858E7	0.886	0.87	0.88	3.14	2.11	2.12
26	230	80	36	1.372/1.	4	2.089E6	2.858E7	0.93	0.926	0.93	3.16	1.687	1.74
27	400	80	72	5.037/1.07	10	4.783E6	16.12E7	0.83	0.878	0.898	3.32	1.19	1.26
28	350	80	72	5.037/1.07	10	4.783E6	16.12E7	0.80	0.846	0.867	3.32	1.23	1.29

of thrust, bending moments, and deflection are then modified by using the factors outlined in Table 5. These modification factors are dependent on the spacing and eccentricity of the stiffeners, rise to span ratio, as well as on the ratio of the rigidity of the stiffened to non-stiffened zones of the conduit walls. It is expected and noticed that the modification factors for bending are almost equal to those calculated for deflection.

ACKNOWLEDGEMENT

The authors acknowledge the contribution by Mr. Osama Elssa who developed a statistical model to determine the modification factors.

NOMENCLATURE

EI_1 unit bending rigidity of non stiffened

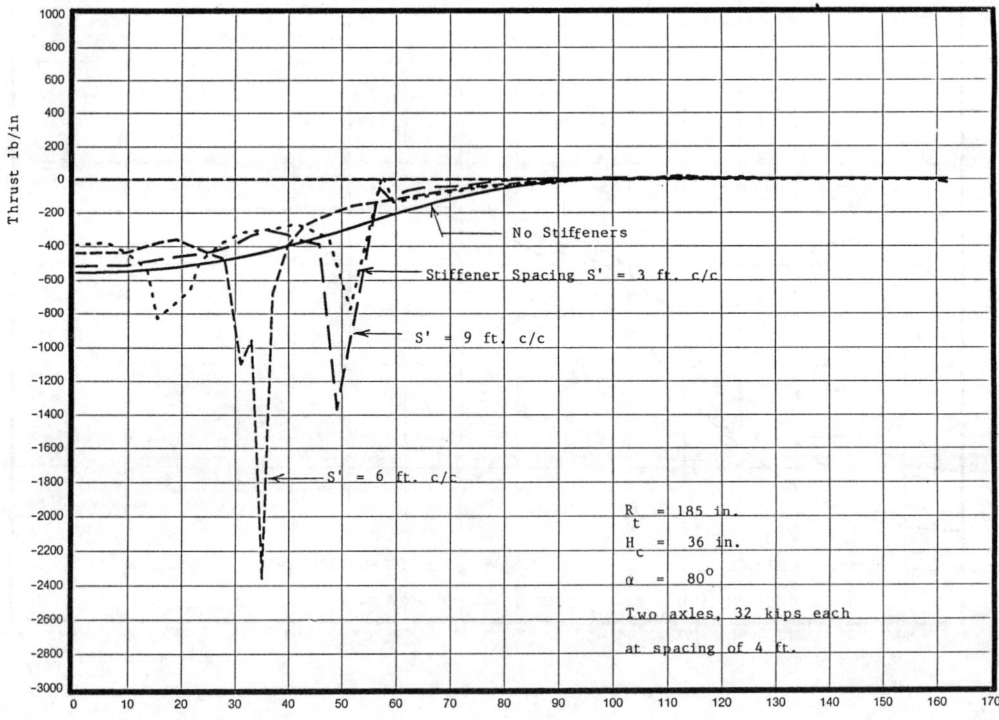

Fig. 4 Thrust at haunch

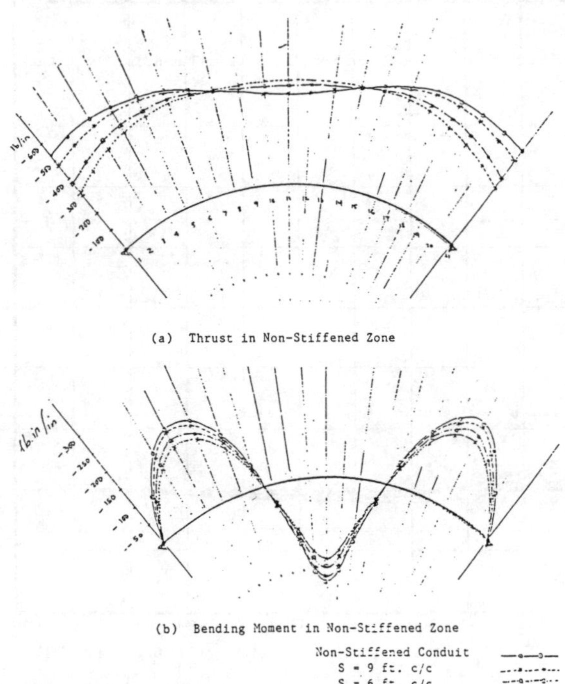

(a) Thrust in Non-Stiffened Zone

(b) Bending Moment in Non-Stiffened Zone

Fig. 5 Effect of stiffeners spacing on thrust and bending moment in
non-stiffened zone

13

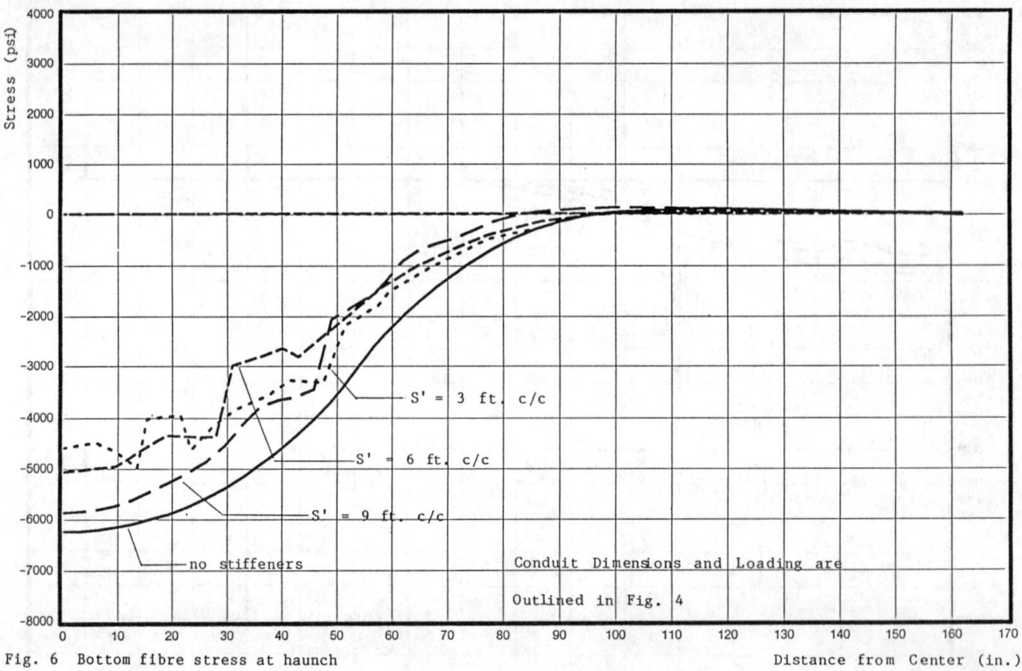

Fig. 6 Bottom fibre stress at haunch

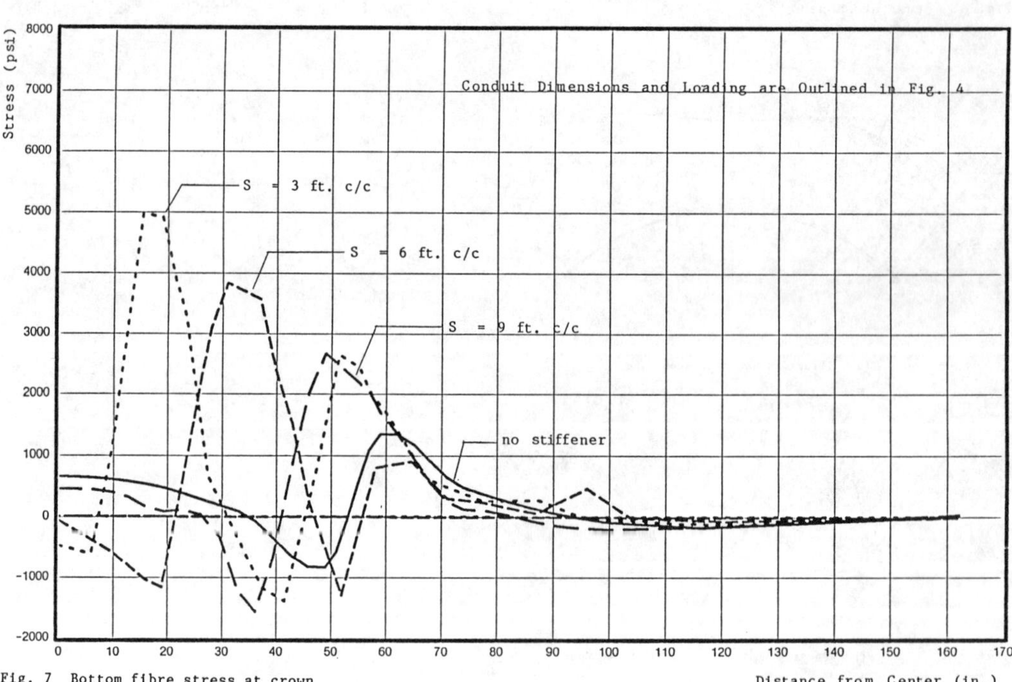

Fig. 7 Bottom fibre stress at crown

14

Table 5. The Modification Factors

$$\gamma = k \left(\frac{S}{R} \right)^a \left(\frac{F}{f} \right)^b \left(\frac{1 - \cos\alpha/2}{2\,\sin\alpha/2} \right)^c \left(\frac{W\,EI_2}{(S-W)\,EI_1 + W\,EI_2} \right)^d$$

Parameter	k	a	b	c	d	Error
Thrust	1.022	0.048	−0.059	0.023	−0.015	±2.22
B. Moment	1.083	0.027	−0.013	0.110	−0.052	±3.55
Deflection	1.068	0.026	−0.017	0.094	−0.042	±3.21
Thrust	0.694	0.399	0.582	−0.806	−1.333	±14.5
B. Moment	0.478	0.580	−0.019	−1.207	−0.722	±13.5
Deflection	0.467	0.534	0.002	−1.184	−0.698	±12.6

zone

EI_2 unit rigidity of stiffened zone
f   half depth of shell corrugation
R   radius of conduit
S   spacing of stiffeners, in.
W   width of stiffeners, in.
α intersection angle of top arch

REFERENCES

Abdel−Sayed, G. 1971. Critical shear loading of curved panels of corrugated sheets. ASCE J. Eng. Mech. Div. 96, No. EM6: 895−912.

Abdel−Sayed, G. and Ekhande, S. 1985. Effect of stiffeners on the behaviour of soil−steel structure. A Final Research Report (IRI−17−1) for the Ontario Ministry of Transportation and Communications, Toronto, Ontario, Nov. 1985.

American Association of State Highway and Transportation Officials 1983. AASHTO Bridge Specification, Washington, D.C.

Cheung, Y.K. 1986. Finite strip method in structural analysis, Pergamon Press.

Ekhande, S.G. 1986. Rib−stiffened corrugated soil−steel structures, Ph.D. Thesis, Department of Civil Engineering, University of Windsor, Windsor, Ontario, Canada.

Ekhande, S. and Abdel−Sayed, G. 1986. Structural performance at rib−stiffened soil−steel structure, Proc. 2nd International Conference on Short and Medium Span Bridges, Ottawa: 359−372.

Ekhande, S. and Abdel−Sayed, G. 1989. Application of compound finite strip method to the 3D analysis of soil steel structures, Canadian Journal of Civil Engineering: 426−458.

OHBDC 1983. Ontario Highway Bridge Design Code. Ontario Ministry of Transportation and Communications, Downsview, Ontario, Canada.

Okeagu, B.N and Abdel−Sayed, G. 1984. Coefficient of soil reaction for buried flexible conduits. ASCE J. Geotech. Eng., Vol. 11, No. 7: 908−922.

Seely, Fred B. and Smith, James O. 1967. Advanced mechanics of materials, 2nd Edition, John Wiley and Sons, Inc., New York.

Structural Performance of Flexible Pipes, Sargand, Mitchell & Hurd (eds) © 1990 Balkema, Rotterdam. ISBN 90 6191 165 6

A comparison of analytical and field data for a rib-reinforced corrugated steel box-type culvert

A.K.Amla, S.M.Sargand & G.A.Hazen
Center for Geotechnical and Groundwater Research, Ohio University, Athens, Ohio, USA

J.O.Hurd
Ohio Department of Transportation, Columbus, Ohio, USA

ABSTRACT: A corrugated steel box culvert with rib stiffeners was instrumented for monitoring strains and deflections. Measurements were recorded during backfilling and under live load. Moments and thrusts were determined from field data with the assumption of non-composite action between the rib and plate.
The Culvert ANalysis and DEsign (CANDE) finite element program was used to analyze the soil-culvert system. The backfill material was simulated with Duncan's hyperbolic model. CANDE analysis was compared with the measured values of deflection, moment, and thrust.

1 INTRODUCTION

Corrugated metal box culverts are increasingly popular substitutes for short span highway and railroad bridges. Box culverts are primarily used in situations requiring a large span with low vertical clearance where circular and elliptical corrugated pipe culverts are unsuitable. Most metal box-type culverts are constructed of corrugated metal plates bolted together in the field to attain the desired shape and length of the culvert. Rib stiffeners, beams, angles or corrugated ribs are usually provided to enhance the flexural rigidity of flexible metal culverts. Several investigators have studied the behavior and applications of box culverts.

Beal (1982) extensively studied aluminum culverts. He concluded that backfill placement distorted the shape of the culvert, that change in height was less than change in rise due to upward movement of the invert, and dead load moments around the circumference were at maximum about 70 percent of the fully plastic value. He also found that stresses due to live loads were small compared to dead load stresses, and design estimates of thrust were greater than measured values while estimates of moment were less. Analysis of a second aluminum culvert by Beal (1986) produced these additional conclusions: slip between plate and stiffening ribs prevents full composite action developing under service loads, deflection and bending increase significantly under additional compaction effort from vehicle loads, finite element analysis adequately predicts bending and crown deflection due to backfilling, and finite element analysis overestimates the effect of live loads.

Selig and Musser (1985) studied the performance of a rib-reinforced long span culvert with field measurement and a limited analysis using the Culvert ANalysis and DEsign (CANDE) program. The researchers found that the circumferential ribs reduced deflections, increased moments, and did not change thrusts. They were also found to help in controlling the shape of the culvert during

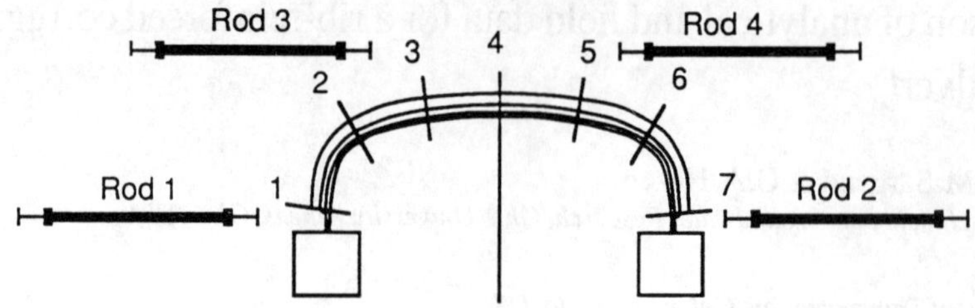

FIGURE 1 LOCATION OF MEASURING INSTRUMENTS

backfilling placement and compaction. Leonards, Juang, Wu, and Stetkar (1985) used CANDE to predict the performance of buried metal conduits. The researchers considered the effects of changing soil models, soil-conduit interface conditions, sequence of soil layer placement, and also the effects of yielding and buckling of the wall. They concluded that the Duncan hyperbolic model adequately characterized soil response and that attention must be paid to the details of backfilling to make valid comparisons with field measurements.

A multi-plate steel box culvert was built and instrumented for this project. Measurements made during the backfilling and under live loads were compared to the results of finite element modeling. The CANDE program developed by Katona, et al. (1976) was used in this study.

2 DESCRIPTION AND INSTRUMENTATION OF CULVERT

The multi-plate corrugated steel box culvert constructed for this investigation had a span of 15 feet 9 inches and a rise of 5 feet. Length of the structure was approximately 40 feet with reinforced concrete headwalls at both ends. Reinforcing rib stiffeners were attached to the outer surface at 2 feet intervals along the length of the structure. The steel angle ribs were 5 X 3 X 0.5 inches in dimension and terminate about 3 inches above the footing. The culvert was built on 3 foot square strip footings. The culvert was built from corrugated steel plate with thickness of 0.2451 inches at the crown and 0.1644 inches on the sides. The corrugations had a 6 inch pitch and 2 inch effective depth. The instrumented cross-section was equidistant from both headwalls.

Deflection data was measured to ±0.015 inches with a tape extensometer at eleven points along the inner circumference of the pipe. Electric strain gauges were mounted at seven evenly spaced locations along the periphery of the culvert shown in Figure 1. Two uniaxial strain gauges were mounted on each of the seven sections, one on the inner surface of the upright leg and the other was mounted on the upper surface of the horizontal leg. Two biaxial strain gauges were attached to the plate at each section, one on the valley of the corrugations and one on the peak. The configuration of strain gauges is illustrated in Figure 2. Soil response was measured at four locations using soil extensometers embedded in the backfill soil at locations shown in Figure 1.

The area surrounding two of the bolts fastening the rib to the plate were instrumented with an array of rosette strain gauges. The first bolt was 6.5 inches above the north foundation, the second 45.1 inches above it. Each bolt had four lines of rosette gauges extending away from the bolt: two of them oriented in the circumferential direction, a third in the longitudinal direction, and the fourth in the diagonal direction.

The live load was applied using a dual-axle dump truck load with limestone to a total weight of 16, 32 and 42 kilopounds, respectively.

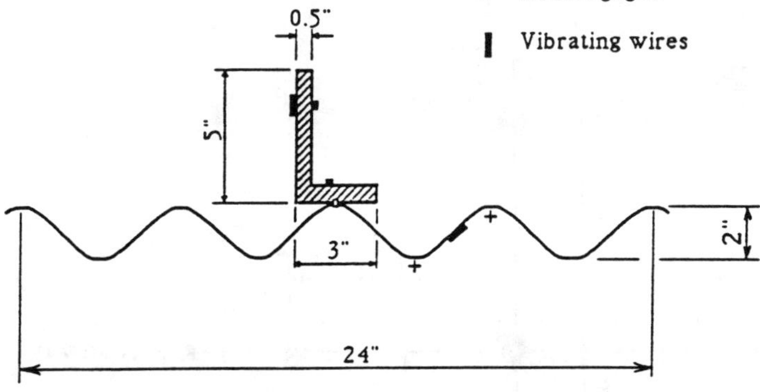

FIGURE 2   TYPICAL INSTRUMENTED SECTION

The load was transferred to tne pavement through the wheels which can be modelled as two point loads along the longitudinal axis of the culvert. Each live load was applied at five positions over the culvert, illustrated in Figure 3.

3 FINITE ELEMENT ANALYSIS

The CANDE program developed by Katona, et al (1976) was utilized for the finite element analysis. A half mesh was used to model the backfill operation and the symmetrical live load positions 1, 3, 4 and 5 in Figure 4.

The linear elastic model was used to represent the concrete footings and asphalt pavement, with parameters obtained from the literature. The Duncan hyperbolic model was used to describe the insitu soil, backfill material and limestone subbase. The parameters for the Duncan model were obtained from extensive laboratory testing; they are listed in Table 1.

Multiaxial and triaxial tests were conducted on the backfill material collected at the construction site (1990). The parameters used in this study are based on these tests. The CTC (Conventional Triaxial Compression) and triaxial tests were used to derive the CTC soil parameters. The general soil parameters were arrived at using all stress paths

(1990). The CTC soil parameters were used to model the backfill sand in regions that were expected to experience compression stress paths such as the soil above the crown. The general parameters were used to model the soil at the sides up to the haunches that experience a wide range of loading paths.

CANDE assumes a plane strain condition and provides only a two dimensional analysis of the system. Duncan's method (1983) was used to convert the point loads into equivalent line loads.

4 RESULTS

Figure 5 illustrates the vertical deflection of the culvert crown during backfill. The crown was moved upward in the initial stages of backfilling, but it was moved downward as backfill material was placed over it. Maximum vertical deflection was measured at the crown after the placement of asphalt. CANDE cannot simulate the stresses from the laying and rolling of asphalt; this partly explains the large discrepancy between theoretical and experimental results. The vertical deflections of the crown due to the live loads at positions 1 and 2 are shown in Figures 6 and 7. The live load deflections are those computed after the backfill, insitu, and foundation deflections have been

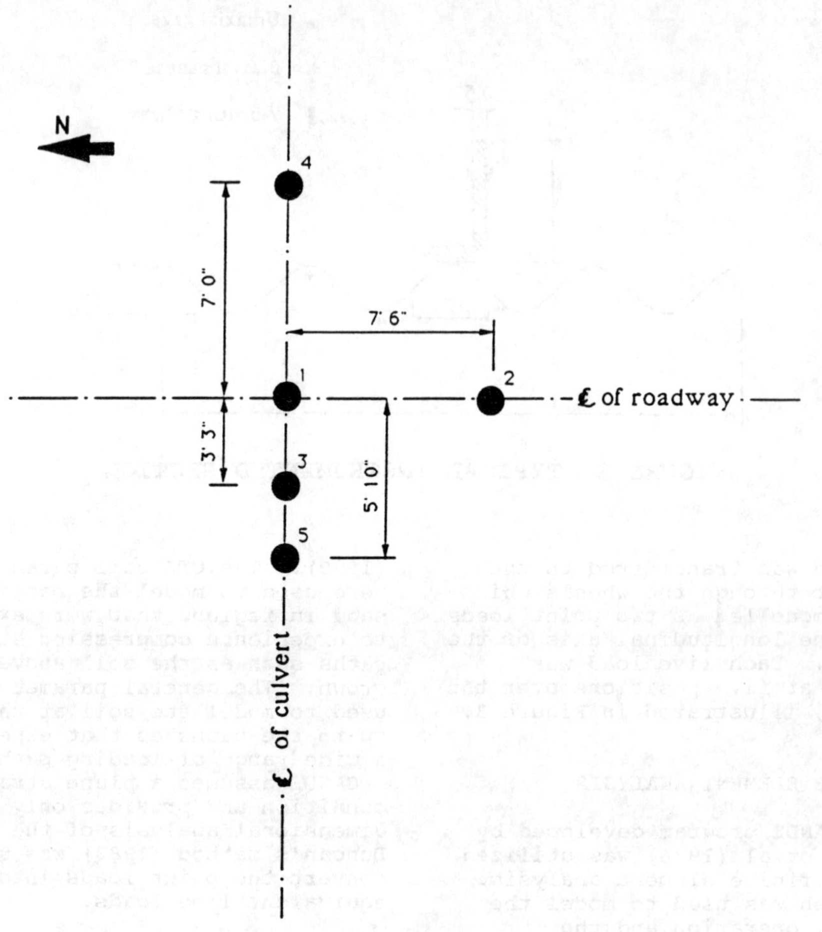

FIGURE 3 LIVE LOAD APPLICATION POSITIONS

subtracted out. The live load deflection is much smaller than the backfill deflection. The CANDE predictions of live load deflection are much larger in magnitude than the actual deflections.

Figures 8 and 9 illustrate the bending moment and the thrust on the culvert after the paving was completed. The finite element and experimental moments match reasonably well, but the thrusts do not. CANDE seriously underestimates the thrusts, and the two sets of maximum thrusts do not occur at the same location, and do not agree in magnitude. This difference in location may be due to the distortion in the culvert

from the uneven backfilling process. The large magnitude of the experimental thrusts indicates that thrust should not be ignored when designing culverts.

Figures 10 and 11 show the moment and thrusts under live load at position 1. Again, the moments correlate very well except at the crown, but the thrusts are underestimated. The crown moment does not compare well because the shear transfer is not sufficient to force a composite response.

The dependence of the analysis on the temperature was studied. Changes in temperature affect the asphalt modulus, which in turn affects the system. Crown

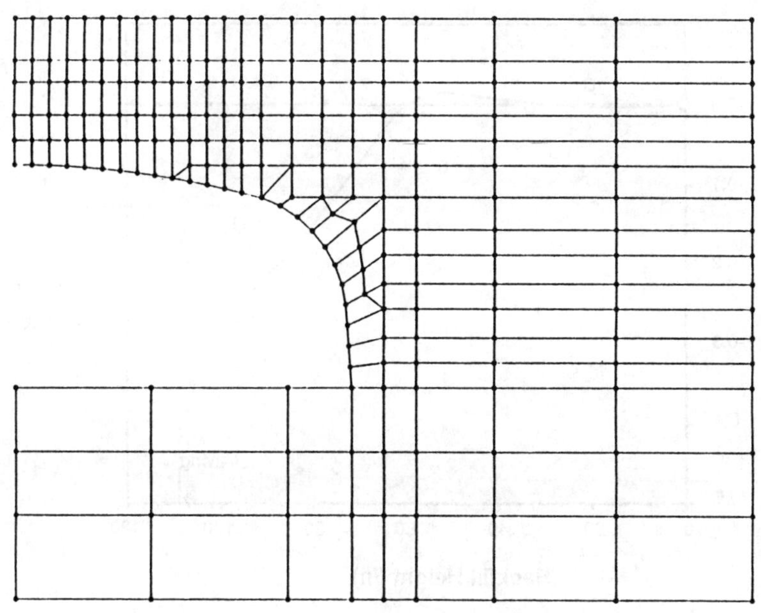

FIGURE 4 TYPICAL CANDE MESH

TABLE 1 PARAMETERS FOR DUNCAN'S HYPERBOLIC SOIL MODEL

Soil Model Parameter	CTC Path Parameters	General Path Parameters
ϕ_0	39 degrees	39 degrees
$\Delta\phi$	5.5 degrees	3.0 degrees
C	0.0	0.0
K	450	1200
n	0.35	1.1
R_f	0.60	0.70
K_b	300	350
m	0.30	0.25

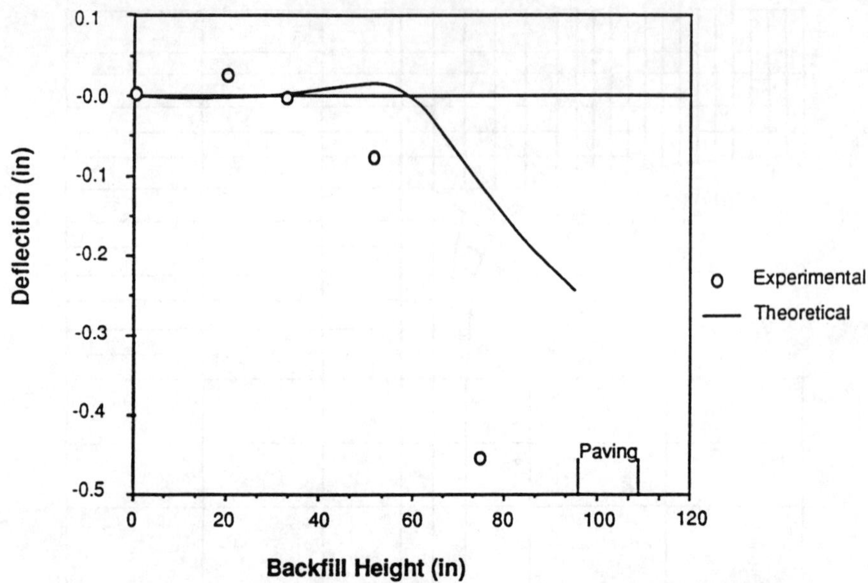

FIGURE 5 VERTICAL CROWN DEFLECTION – BACKFILL HEIGHT CURVE

deflection increased by approximately 7 percent for each 5 degree F increase in temperature under a 42 kip load. Moment, on the other hand, increased by about 10 percent with each 5 degree F increase under the 42 kip load. It is thus important that temperature during the live load be noted.

5 CONCLUSIONS

The conclusions reached from the comparison of field results and CANDE predictions are:
1. For well controlled backfill procedures, the deflections in the culvert are reasonable. The moments are within design limits. However, the thrusts must be considered in the design procedure by reducing the plastic moment capacity to allow for thrust loading capacity. This finding contradicts previous investigations.
2. CANDE predicts moments and deflections during backfill and live load with reasonable accuracy. During paving, the culvert experienced loading and unloading. The CANDE computer program does not include a material mode that can simulate the loading and unloading that took place during pavement placement operations.
3. A large portion of the pavement deformation took place during paving operations.
4. Thrust measurements and the corresponding CANDE results do not correlate at all. CANDE underestimates the thrust by a large amount. The experimental thrusts were asymmetric and of a different magnitude.

REFERENCES

Beal, D.B. 1982. Behavior of aluminum structural plate culvert. Transportation Research Board. 878:100-104.
Beal, D.B. 1986. Behavior of corrugated metal box culvert. Federal Highway Administration, Department of Transportation. Report No. RR-86-133.
Selig, E.T. & Musser, S.C. 1985. Performance evaluation of a rib-reinforced culvert. Transportation Research Record. 1008:117-122.
Leonards, G.A., Juang, C.H., Wu, T.H. & Stetkar, R.E. 1985.

Predicting performance of buried metal conduits. Transportation Research Record. 1008:45-52.

Katona, M.G. & Smith, J.M. 1976. CANDE: User Manual. Federal Highway Administration. Report No. FHWA-RD-77-6.

Rauch, A.F. 1990. Experimental and numerical investigtion of a deep-corrugated steel, box-type culvert. Master's thesis presented to the Faculty of the College of Engineering and Technology, Ohio University.

Duncan, J.M. & Drawsky, R.H. 1983. Design and performance of aluminum box culverts. Dept. of Civil Engineering, University of California, Berkeley. Report No. UCB/GT/83-04.

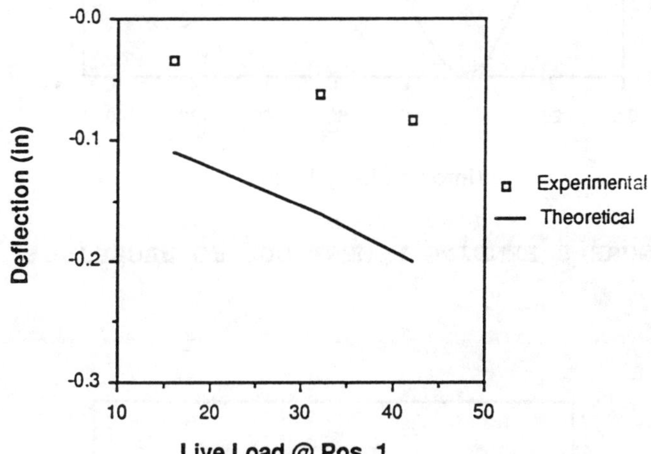

FIGURE 6 VERTICAL CROWN DEFLECTION DURING LIVE LOAD TESTS AT POSITION 1

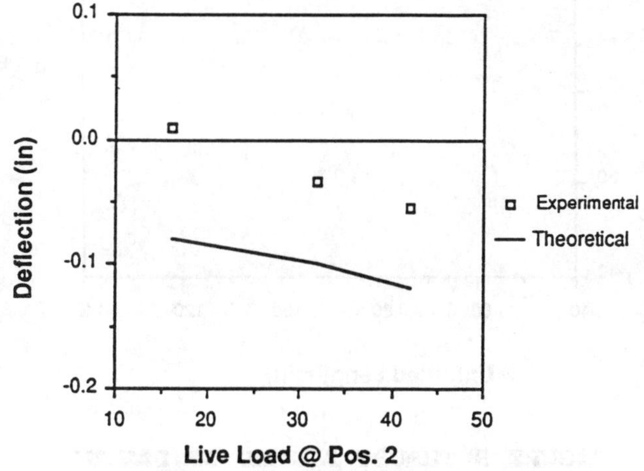

FIGURE 7 VERTICAL CROWN DEFLECTION DURING LIVE LOAD TESTS AT POSITION 2

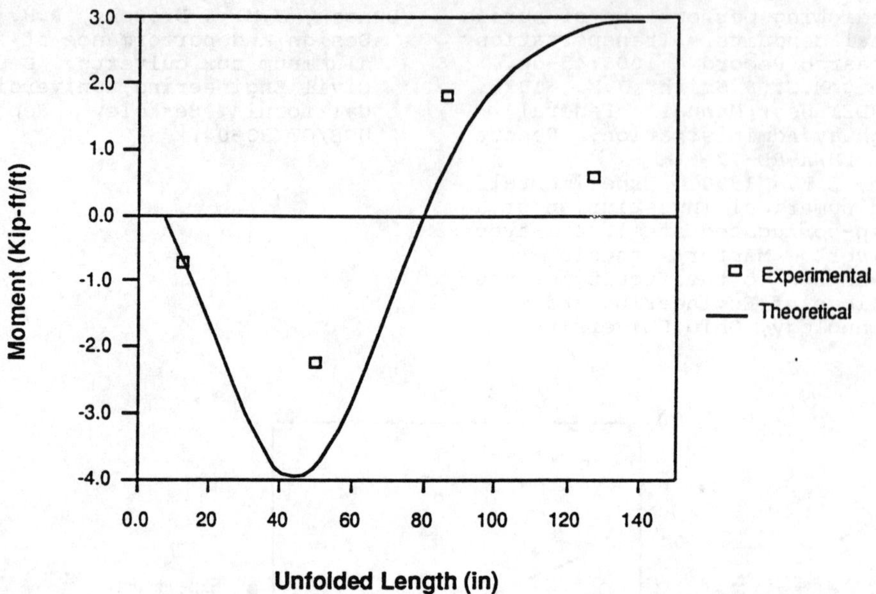

FIGURE 8 BENDING MOMENT DUE TO BACKFILL

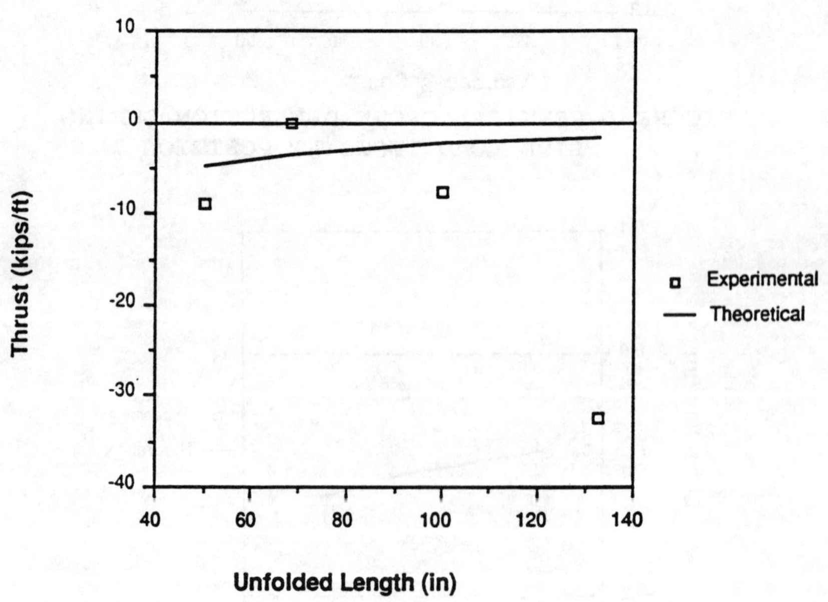

FIGURE 9 THRUST AT END OF PAVING

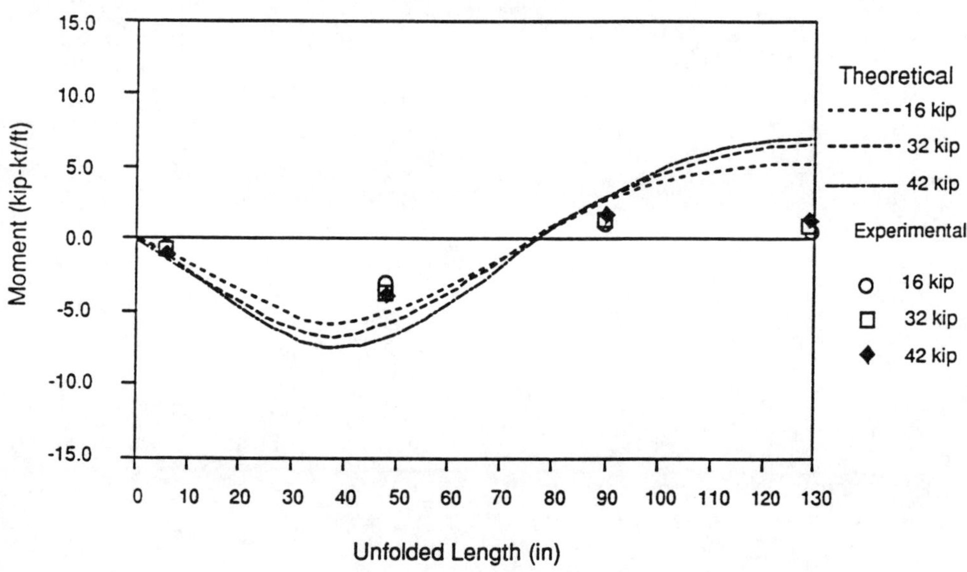

FIGURE 10 BENDING MOMENT DUE TO LIVE LOADS

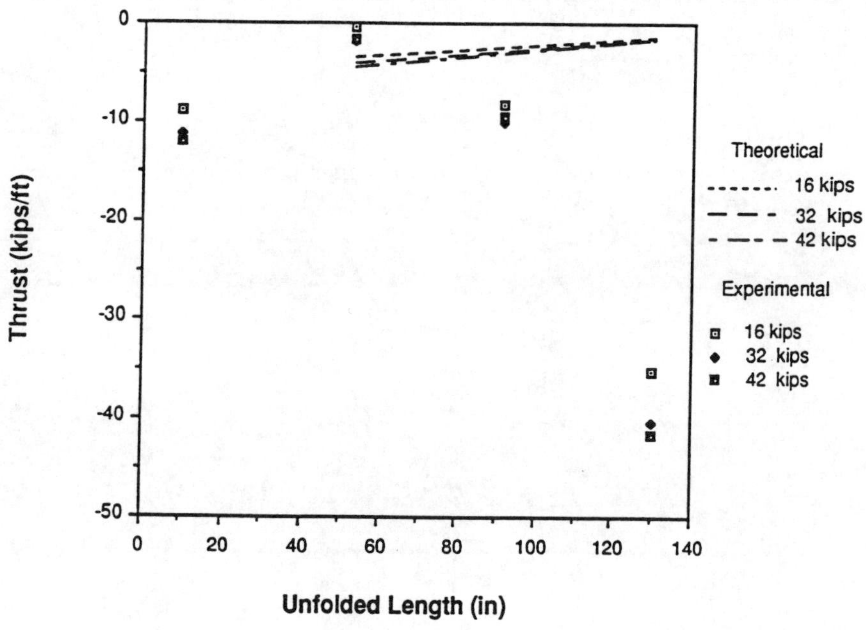

FIGURE 11 THRUST – UNFOLDED LENGTH CURVES
UNDER LIVE LOADS

Structural Performance of Flexible Pipes, Sargand, Mitchell & Hurd (eds) © 1990 Balkema, Rotterdam. ISBN 90 6191 165 6

Field measurements and analysis of a large diameter flexible culvert

P.M.Byrne & T.Srithar
Department of Civil Engineering, University of British Columbia, Vancouver, B.C., Canada

C.B.Kern
Ministry of Transportation and Highways, Province of British Columbia, Victoria, B.C., Canada

ABSTRACT: This paper describes field measurements and analysis carried out on the Elkhart Creek soil arch culvert structure in British Columbia, Canada. The structure has a span of 13.4 m, a rise of 7.3 m, and a soil cover of 9.6 m. The original structure collapsed during backfilling in October 1987. A new structure of the same design was built in the Fall of 1989, and because of controversy regarding the design thrust value, it was instrumented with 3 "rings" of strain gauges comprising 96 gauges in all. In addition, displacements of the arch were observed during and after construction. Displacements and stresses in the soil were also measured.

The measured thrust values were much lower than expected and indicated that significant positive soil-arching occurred, similar to that observed at the Vieux Comptoir soil-structure in Quebec in 1975.

A nonlinear finite element analysis of the soil-structure system was carried out simulating the construction procedures used, and the computed response compared with the measurements. The computed and observed responses were in reasonable agreement in all aspects; thrust, displacements and soil stresses. The important factors in the analyses are the relative stiffness of the metal arch and soil components of the soil-structure system. In this regard it was very important to allow for slip at the bolts thus greatly reducing the stiffness of the metal arch. The analyses showed that very significant positive arching occurred in agreement with the measurements when such slip was considered.

1 INTRODUCTION

The Elkhart Creek soil-arch structure is a long-span high cover culvert located on the Coquihalla Highway in the interior of British Columbia, Canada. The original culvert collapsed during construction in October 1987 when the soil cover was about 1 m. above the crown.

A new structure of identical design and using many of the original steel plates and the same foundations was constructed during the period August 1 to October 5, 1989. Because of a concern regarding the magnitude of the thrust in the metal arch under such a high backfill, the structure was instrumented with 3 "rings" of strain gauges to measure the thrust. Displacements of the arch were also monitored. In addition, load cells and displacement transducers were placed in the soil fill to help obtain a better understanding of the interaction between the soil and the metal arch during construction.

This paper describes these measurements. In addition, a stress and deformation analysis of the structure was carried out using finite elements to model the soil, and structural beam-column members to model the steel members. This analysis allows the simulation of the construction procedure as well as the nonlinearities of both the soil and structural elements. The predictions from the analysis are compared with the field measurements.

2 DESCRIPTION OF THE CULVERT

A cross-section of the soil-arch metal culvert is shown in Fig. 1. The culvert has a span of 13.4 m, a rise of 7.3 m above the foundation level, and a soil cover of 9.6 m as shown in the figure. The concrete foundations for the arch are founded on rock or very stiff soil. The structure forms an underpass beneath the 4 lane Coquihalla highway with a channel to

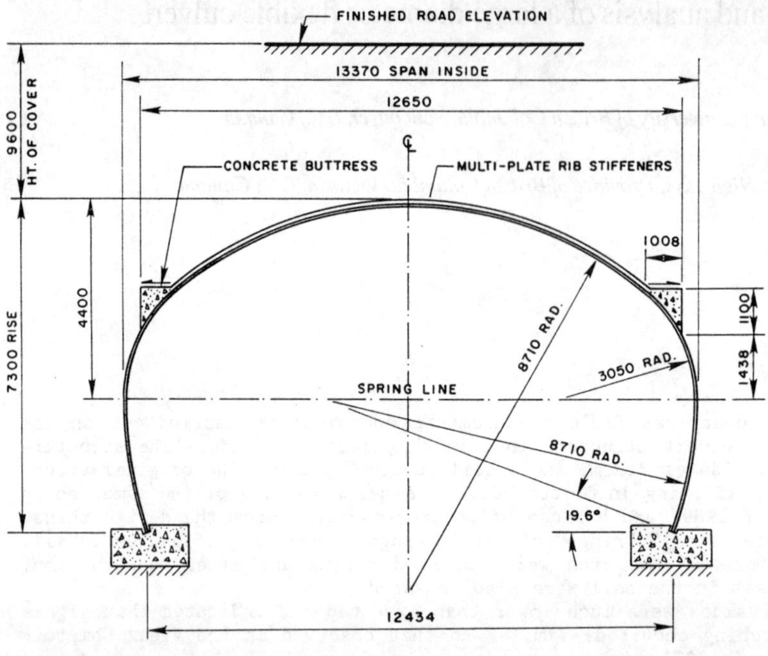

Fig. 1. Culvert cross-section.

Table 1

Plate Thickness (mm)	Area (mm²/mm)	Moment of Inertia (mm⁴/mm)	Section Modulus (mm³/mm)
7.0	8.712	2675.11	92.56

carry a small stream and a path to allow passage of animals beneath the highway (Fig. 2). The axis of the structure is perpendicular to the highway and the structure has a length of 78 m.

The original structure which failed during construction in 1987 was dismantled and the metal arch rebuilt in August 1989 to the same design and located on the same foundations. The arch comprises 7 mm (0.276") corrugated steel plate with the properties shown in Table 1.

Rib stiffeners were attached to the upper part of the arch extending from the concrete thrust beam and over the crown as shown on Fig. 1. These ribs comprised 0.61 m wide strips of the same 7 mm corrugated plate bolted to the arch and located at 3 m spacing.

The plates were bolted together with 19 mm (3/4") diameter bolts inserted in 24 mm (15/16") holes. A section of the arch comprises of 16 lines of holes (including the bolts to the foundation bracket). If it is assumed that the plate holes were initially lined up and that the bolt were centered in the hole, then the amount of relative slip that could later take place between the plates is the difference in diameter between the bolt and the hole, namely 5 mm (3/16"). Since there are 16 such locations of possible slip the total amount of arch shortening that could occur due to slippage is 76 mm (3").

The backfill comprised of natural sand and gravel compacted to between 95% and 100% of standard proctor density. The material at this density had a unit weight of 22.7 kN/m³ (144 lbs/ft³). A loose uncompacted cushion of sand was placed above the crown with an appoximate unit weight of 16 kN/m³ (102 lbs/ft³).

The construction sequence was as follows:

1) The arch was bolted together in place with the bolts tightened to a torque of 200 N.m to 340 N.m.

2) Granular backfill was placed from either side and compacted. The difference in level from side to side was restricted

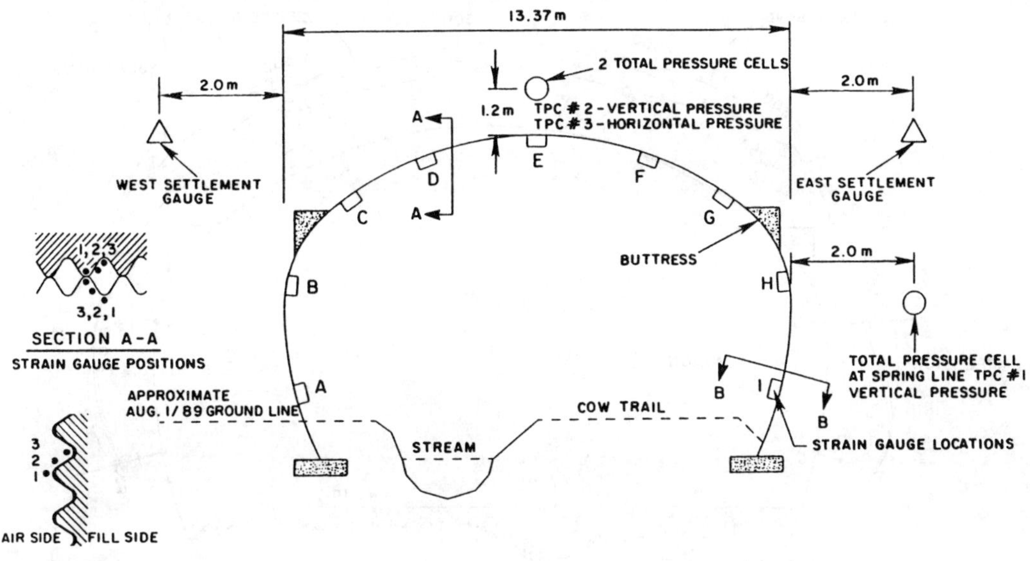

Fig. 2. Instrumentation.

to less than 0.3 m (1 ft.).

3) When the level of backfill reached the elevation of the concrete thrust beams, these were cast.

4) The sand cushion comprising 1.0 m of loose uncompacted sand was then placed over the crown.

5) Backfilling from either side was then continued until the backfill over the crown reached a height of 2 m. During this stage, great care was taken to insure that heavy equipment did not get too close to the metal arch.

6) Once the backfill exceeded a height of 2 m above the crown, heavy equipment was allowed to cross the crown and backfilling continued until the full 9.6 m height of backfill above the crown was achieved.

3 INSTRUMENTATION AND MEASUREMENT

The instrumentation comprised of:

1) Strain gauges on the metal culvert allowing axial and bending strains to be measured;

2) Reference points on the culvert and a fixed reference point on the floor allowing the absolute displacements of the culvert to be measured;

3) Earth pressure cells allowing the total stress in the soil to be measured; and

4) Displacement gauges in the soil.

The strain gauges, earth pressure cells, and displacement gauges were mounted and monitored by Weir-Jones Engineering Consultants.

4 STRAIN GAUGES

Three "rings" of strain gauges comprising a total of 96 gauges in all were used. Each "ring" of gauges was located on a cross-section and comprised of 9 measurement points (A to I) located around the arch as shown in Fig. 2. Each measurement point comprised of 3 strain gauges. One in the hollow of the corrugation, one on the hump and one in between at the neutral axis of the section as shown in Fig. 2. The axial strain at any location was taken as the average of the 3 strains. A "ring" of strain gauges therefore comprised of 9(3) = 27 gauges and the 3 rings comprised of 81 gauges. An aditional 15 gauges were placed on a rib section. The 3 rings were located near the centre of the culvert approximately 4.6 m apart and are referred to as the north (N), centre (C), and south (S) rings. The centre ring contained the rib section.

The measured axial strains at the 3 sections at the end of backfill are shown in Fig. 3. The maximum axial strain is 506 Microns compared with a yield strain

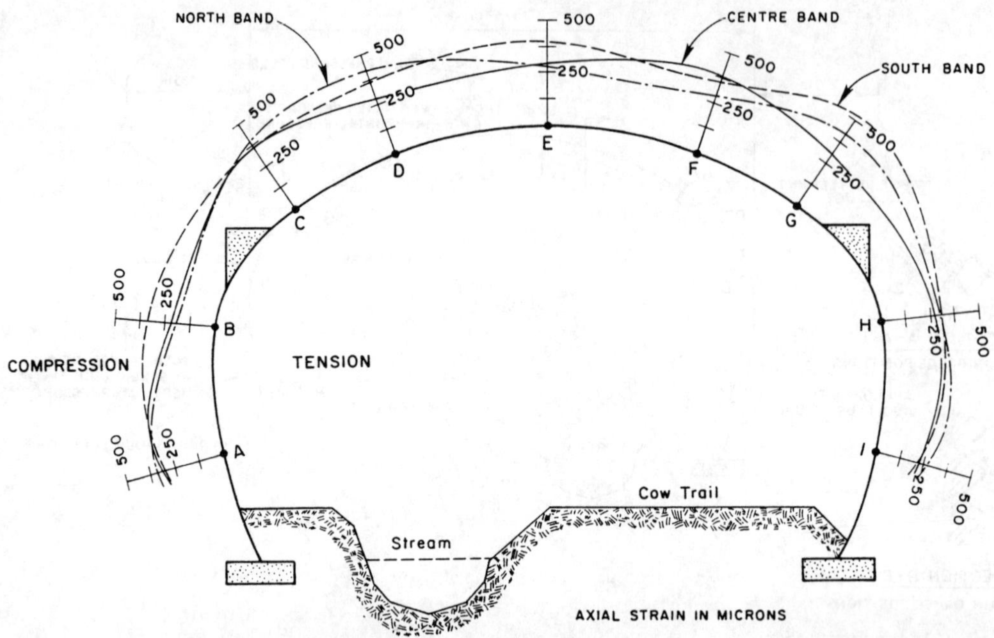

Fig. 3. Measured axial strains in the metal arch.

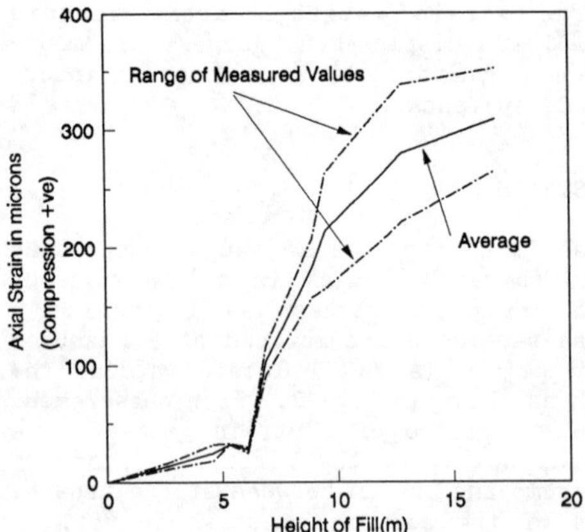

Fig. 4. Axial strain at the crown vs
height of fill above foundation
level.

for the steel of 1138 Microns. The
measured strain implies a maximum axial
stress at the centroid of the section of
101 MPa and compares with yield stress
227.5 MPa. The measured axial stress is
considerably lower than computed by
formulae suggested by Duncan (1) which
gave a value of 272 MPa. This is in
excess of the yield stress. His formulae
imply negative arching. However, even

without arching the axial stress would be
193 MPa. This value is still almost twice
the measured value and implies that positive arching occurred. The low values of
the measured axial strains are in agreement with those measured at the Vieux
Comptoir soil-arch structure in Quebec in
1975 (2). This structure is very similar
to the Elkhart Creek structure and the
maximum measured axial strains there were
also about 500 Microns.

The average axial strain at the crown
as a function of the height of backfill
above the foundation is shown in Fig. 4.
It may be seen that the axial strains
build up rapidly as the fill height is
increased from about 6 m to 13 m, and more
slowly thereafter.

5 CULVERT DISPLACEMENTS

Displacements of the metal culvert during
construction were observed by measuring
the change in distance between reference
points on the culvert and a fixed reference point at the base of the structure.
The points are shown in Fig. 5. Measurements were made at 13 cross-section
locations within the culvert. The movements of the crown, Point 4, and spring
line Point 1, as functions of the height
of backfill are shown in Fig. 6. It may

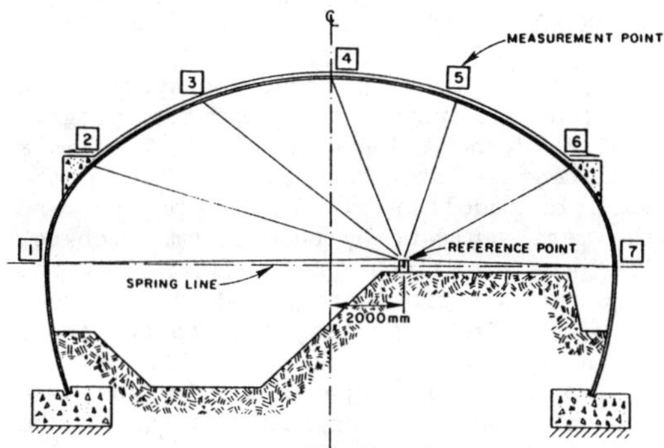

Fig. 5. Culvert displacement measurement points.

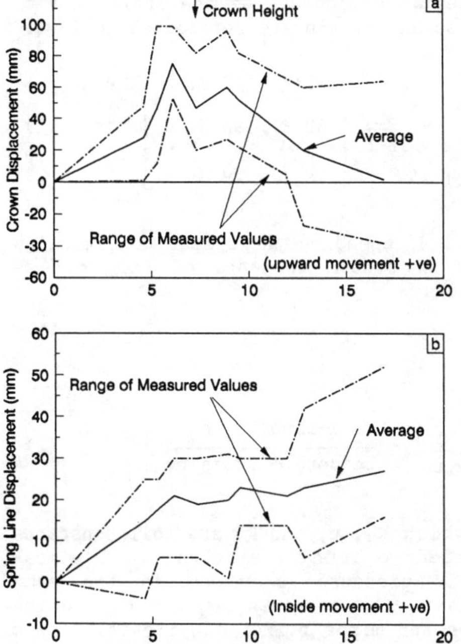

Fig. 6. Culvert displacement vs height of fill above the foundation level.

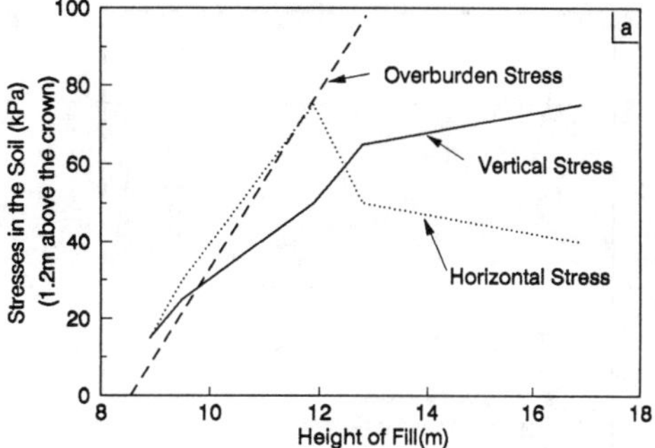

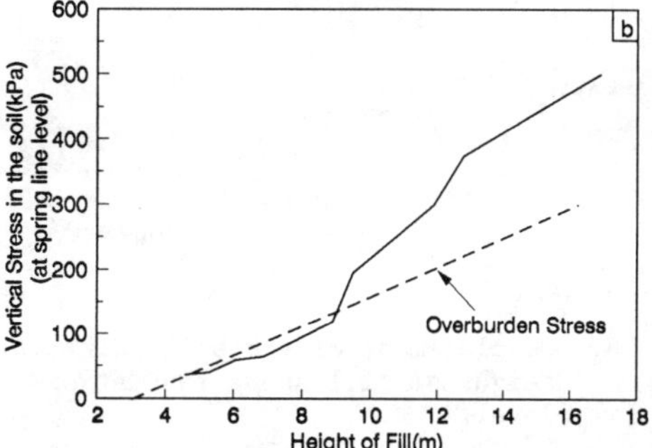

Fig. 7. Earth pressure vs height of fill above the foundation level.

be seen that the crown moved upwards an average of 75 mm until the cushion was placed and thereafter generally moved downwards a similar amount. The spring line point moved inwards at all stages of loading with an average maximum inward movement of about 25 mm.

6 EARTH PRESSURES

Earth pressure cells were located above the crown at a height of 1.2 m. Two cells were placed, one horizontally and one vertically so as to measure both the vertical and horizontal stresses. An earth pressure cell was also placed at the spring line elevation at a distance of 2 m from the culvert (Fig. 2). At this location only the vertical stress was measured.

The measured stresses as a function of the height of fill above the foundation level are shown in Fig. 7. The results indicate that the stresses above the crown (7a) are lower than the corresponding stresses from the weight of the overlying column of soil (overburden stress), whereas adjacent to the spring line (7b), the reverse is true. This indicates that arching is transferring the soil load from the metal arch and to the soil at the sides of the arch. This implies that the surrounding soil is stiffer in the vertical direction than is the arch and is causing positive arching.

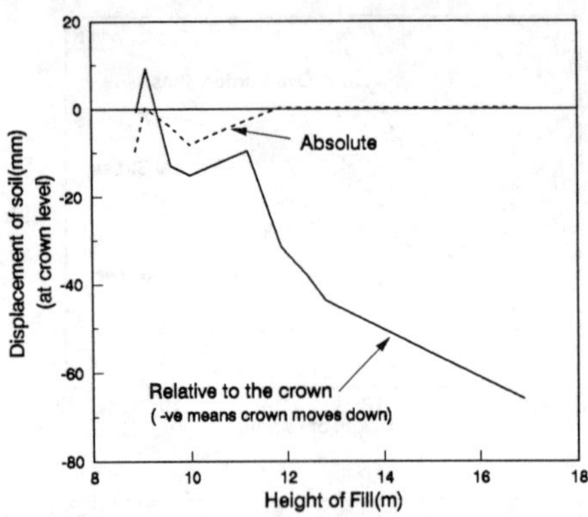

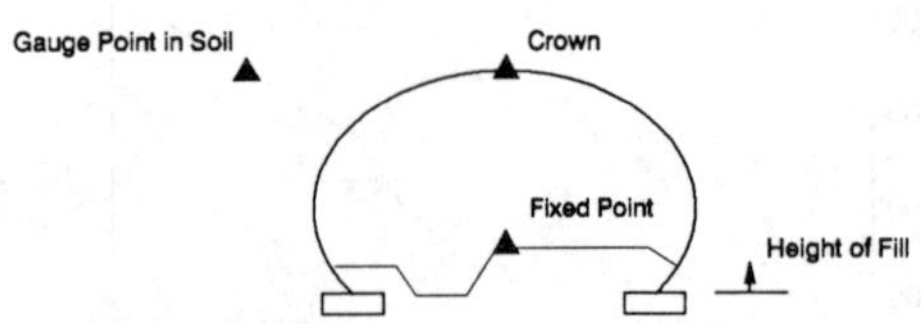

Fig. 8. Displacement of the soil point vs
height of fill above foundation
level.

7 SOIL DISPLACEMENTS

Settlement gauges to measure the vertical
displacements of the soil adjacent to the
crown were installed. Each instrument
comprised of two boxes filled with oil and
connected by a tube. One box was placed
on the crown of the metal arch while the
other was placed in the soil at the level
of the crown but offset 8.7 m perpendicular to the culvert as shown in Fig. 2. By
monitoring the difference in oil pressure
in the two boxes, the differential movement between the crown and the point in
the soil is obtained. The absolute movement of the soil is obtained by adding the
absolute movement of the crown to the
relative displacement obtained from the
gauge. Two of these gauges were deployed
but only one worked.

The measured relative and absolute
displacements of the soil at the gauge
point as a function of fill height above
them are shown in Fig. 8. The relative
displacements indicate that the crown
moved down 66 mm compared with the soil,
indicating a positive arching situation in
agreement with the findings from the
strain gauges and the load cells. The
absolute displacements indicate that the
maximum downward movement of the soil at
the offset point was about 10 mm.

8 METHOD OF ANALYSIS

Analyses of the soil-structure system were
carried out using the computer program
NLSSIP (Byrne and Duncan (3)). This is a
nonlinear elastic program in which the
soil is modelled by finite elements and
the metal culvert by beam-column members.
The program allows the simulation of the
construction procedure in which the soil
is placed in layers adjacent to and above
the structure.

The nonlinear aspects of the soil are
incorporated by considering it to be
incremental elastic and isotropic having
two elastic parameters: a tangent Young's
modulus, E, and a tangent bulk modulus, B,
both of which depend on the current stress
conditions and change at each step of
loading. For plane strain conditions the
cartesian components of the increments of
stress and strain are related as follows:

$$
\begin{Bmatrix} \Delta\sigma_x \\ \Delta\sigma_y \\ \Delta\tau_{xy} \end{Bmatrix} = \frac{3B}{9B-E} \begin{bmatrix} (3B+E) & (3B-E) & 0 \\ (3B-E) & (3B+E) & 0 \\ 0 & 0 & E \end{bmatrix} \begin{Bmatrix} \Delta\epsilon_x \\ \Delta\epsilon_y \\ \Delta\gamma_{xy} \end{Bmatrix} \quad (1)
$$

E and B depend on the stress condition as
well as other constants of the soil as
follows:

$$
E = K_E \cdot p_a \left(\frac{\sigma_3}{p_a}\right)^n
\left[1 - \frac{R_f(1-\sin\phi)(\sigma_1-\sigma_3)}{2c\cdot\cos\phi + 2\sigma_3\cdot\sin\phi} \right] \quad (2)
$$

in which K_E, n, and R_f are soil constants
related to Young's modulus, p_a is atmos-
pheric pressure expressed in the same
units as the stresses, σ_1 and σ_3 are the
major and minor principal stresses, and c
and ϕ are the Mohr-Coulomb strength parameters. Duncan et al. (4) developed the
following expression for B:

$$
B = K_b p_a \left(\frac{\sigma_3}{p_a}\right)^m \quad (3)
$$

in which K_b and m are soil constants
related to the bulk modulus.

The methods for obtaining the parameters associated with both E and B from
triaxial tests are described in detail by
Duncan and Chan (5) and Duncan et al.
(4).

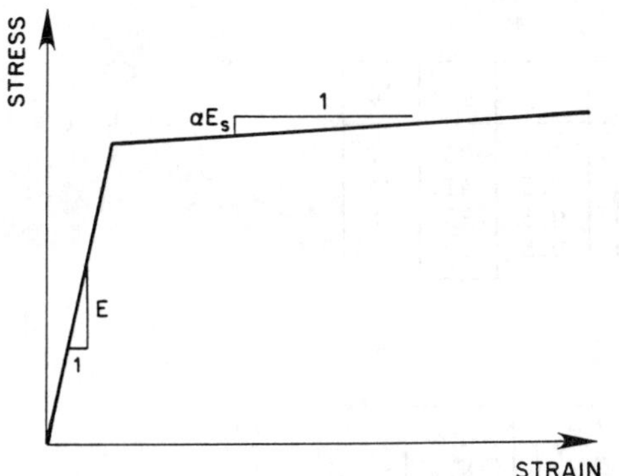

Fig. 9. Assumed stress-strain relations for beam members.

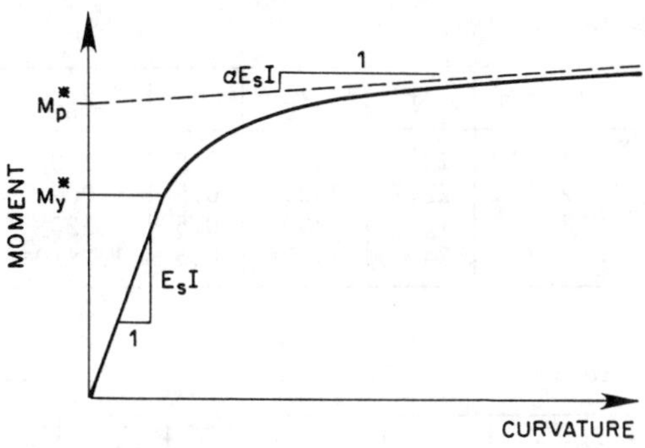

Fig. 10. Moment-curvature relations for beam members.

The arch structure is considered to comprise of a collection of interconnected straight beam members which under load may be subjected to both axial force and bending moment. The beams may behave in a nonlinear fashion due to (a) the nonlinear nature of the stress-strain relations of the material at high stress levels, and (b) significant changes in geometry. The stress-strain relations for the beam materials are assumed to be bilinear as shown in Fig. 9. The parameters used to define the lines are: E_s = the slope of the stress-strain curve of the steel in the elastic range, F_y = the yield stress, α = the ratio of the slope of the stress-strain curve in the plastic range to that in the elastic range.

The stress-strain relationship is used to develop a moment curvature relationship such as that shown in Fig. 10. Both the moment at which yielding first occurs, M_y^*, and the moment at which the section becomes fully plastic, M_p^*, depend on the magnitude of the axial force, P, and are given by:

$$M_y^* = M_y (1 - P/P_p) \qquad (4)$$

$$M_p^* = M_p (1 - (P/P_p)^2) \qquad (5)$$

in which M_y and M_p are the yield and plastic moments if the axial force is zero, and P_p is the fully plastic axial force when the moment is zero. Equation 5 is correct for members having a rectangular cross-section and is a good approximation for corrugated sections. The terms

in the incremental stiffness matrix that arise from flexure are derived from the slope of the moment curvature relationship. The moment at either end of the beam will be different and consequently a variation in slope will exist along the member and this is accounted for as described by reference (3).

Geometry changes can be accounted for by satisfying equilibrium of the system in its displaced position under the total load. However, an incremental solution technique is required for the soil elements and consequently an incremental solution technique must be used for the beams. An approximate incremental solution is obtained by upgrading the coordinates of the nodes of the beam members after each increment of load and by modifying the member stiffness matrix to include the effect of the axial force.

9 MODELLING OF THE SOIL-STRUCTURE SYSTEM

The finite element and structural modelling of the soil-structure system is shown in Fig. 11. Only half of the system was modelled as the geometry and loading are essentially symmetric. The arch structure was modelled by 8 beam-column members and the soil by 150 finite elements placed in 11 layers. The concrete foundation strip was also modelled by finite elements. The properties used for the various backfill soils are listed in Table 2.

These are based on laboratory and field experience with similar soils from Duncan et al. (4) and Byrne et al. (6).

The properties used for the beam elements are listed in Table 3.

The ribs comprised of 0.61 m wide strips of the same corrugated plate bolted

Table 2

Soil Type	γ	K_E	n	K_B	m	K_o	ϕ	$\Delta\phi$
1	22.7	600	0.5	360	0.25	0.5	36	2
2	22.7	2000	0.5	1000	0.25	0.5	41	4
3	16	200	0.5	120	0.25	0.5	33	0
4	24	2×10^5	0.5	1.2×10^5	0.25	0.5	60	0

Table 3

Elements	E_s (kPa)	I (m^4/m)	A (m^2/m)	α	β	P_y (kN)	M_y (kN.m)	M_p (kN.m)
without ribs	$2(10^8)$	2.675 (10^{-6})	8.712 (10^{-3})	0.0067	1.0	1982	23.9	37.7
with rib stiffeners	$2(10^8)$	3.799 (10^{-6})	10.45 (10^{-3})	0.0067	1.0	2378	34.0	53.6

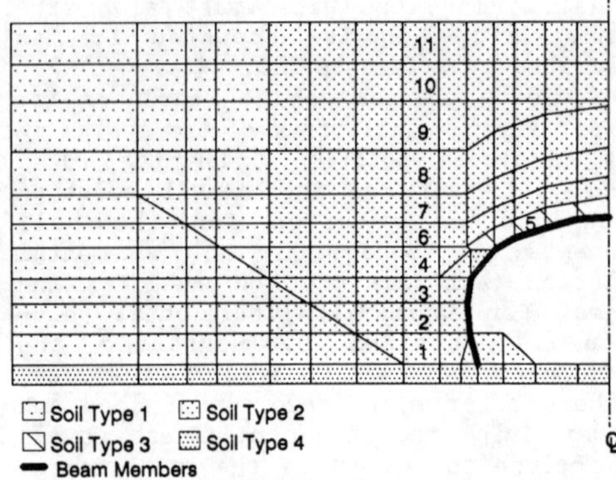

Fig. 11 Numerical modelling of arch and soil elements.

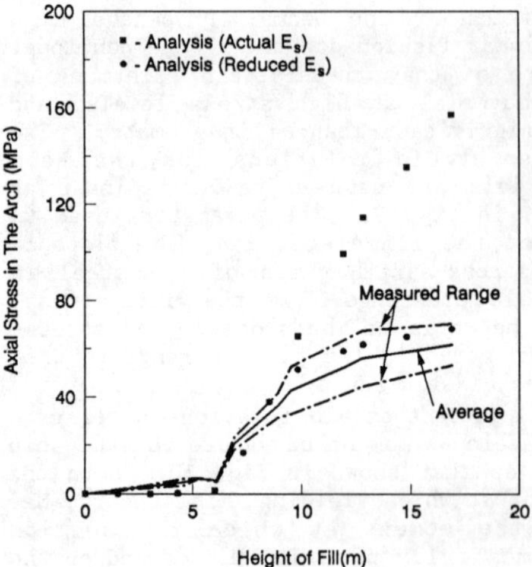

Fig. 12. Axial stress in the arch at the crown vs height of fill.

to the arch at 3 m spacing. The equivalent properties of the stiffened sections were determined by computing the combined properties for a 3 m length and then dividing by 3.

10 RESULTS

The predicted axial stress in the arch at the level of crown as a function of the height of fill above the footing level is shown as the solid square symbols in Fig. 12. Also shown on this figure are the measured range of axial stress values together with the average values. As the fill height increases the measured and

predicted axial stresses are in reasonable agreement. However, when the fill height exceeds about 10 m above the footing (3m above the crown), further increase in measured stress is small, while the predicted stress continues to increase significantly. It is thought that this sudden drop in the measured buildup of thrust is due to slippage at the bolted connections causing a drop in the effective stiffness of the arch. This slippage was accounted for in the analysis by reducing the axial stiffness of the

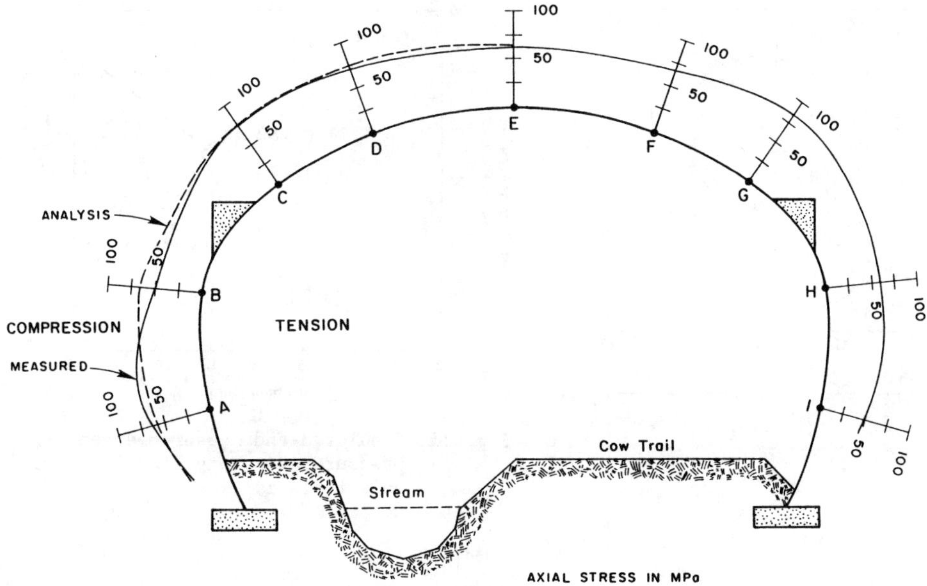

Fig. 13. Predicted and measured axial stress distribution in the arch.

structural members while maintaining their bending stiffness. A reduction in stiffness by a factor of 50 was found by trial to predict results in reasonable agreement with the measurements as shown by solid circles in Fig. 12. The predicted shortening of the arch due to the reduced stiffness was 50 mm which is below the 76 mm of potential slip computed earlier.

Unfortunately, no measurements of slippage at the bolts were made. However, it was found that the sudden change in the rate of buildup of thrust in the arch could not reasonably be accounted for in any other way. In addition, as will be shown next, other predictions and measurements are in agreement with the concept that a sudden drop in effective arch stiffness occurred when the fill reached a height of 10 m above the footing (3m above the crown).

The measured and computed axial stress distributions in the metal arch are shown in Fig. 13 and are seen to be in close agreement. The computed values shown have a 50 fold reduction in E_s to allow for slippage when the fill height exceeds 3 m above the crown. Without such a reduction the computed values are about 2.5 times higher, which is in close agreement with the prediction from Duncan's equations discussed earlier.

Analyses for two conditions were carried out.

Analysis 1: In which the axial stiffness of the arch is based on the E_s of steel at all stages of loading; and

Analysis 2: In which the axial stiffness of the arch is based on E_s until the fill height reaches 10 m. Thereafter the axial stiffness is reduced by a factor of 50 while maintaining the bending stiffness.

However, for the sake of clarity only the Analysis 2 results will be shown on subsequent figures.

The predicted and measured stresses in the soil at the load cell location 1.2 m above the crown are shown in Fig. 14. It may be seen that predicted and measured vertical stresses are in reasonable agreement and are well below the overburden stress, indicating positive arching (Fig. 14a). The predicted horizontal stresses are not in good agreement with the measured values, being significantly higher (Fig. 14b). However the analyses indicated that the horizontal stresses are very sensitive to location. In the cushion just below the gauge point the predicted horizontal stresses are much lower than the measured values shown.

The predicted and measured vertical stresses in the soil adjacent to the spring line as a function of the fill height are shown in Fig. 15. It may be seen that the predicted values are in very good agreement with the measured ones. Also shown is the overburden stress, and it may be seen that the measured vertical stresses are higher than the overburden stress, indicating that positive arching

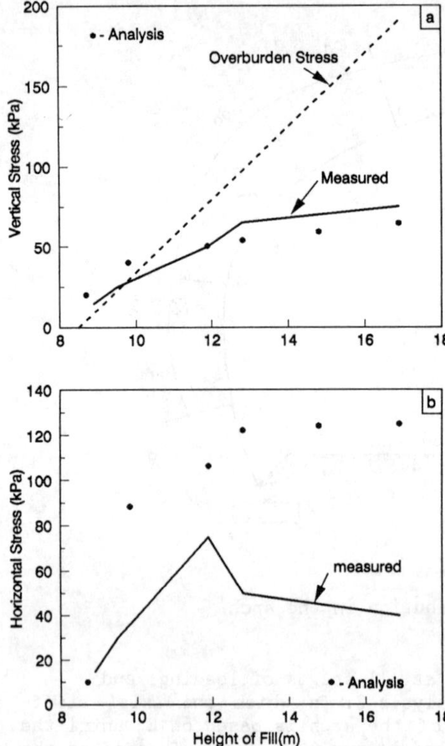

Fig. 14. Measured and predicted earth
pressures at 1.2 m above the
crown vs height of fill.

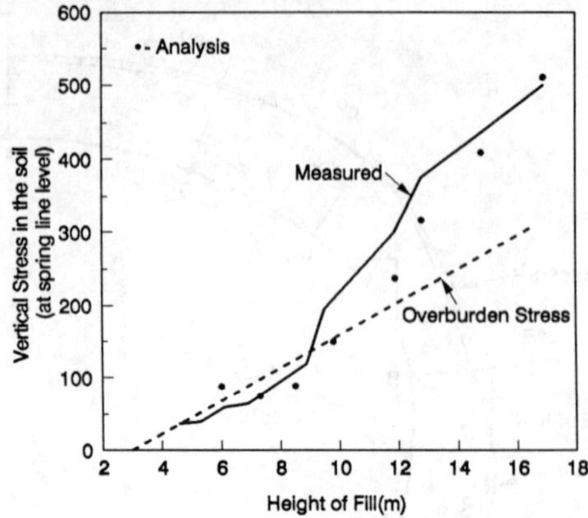

Fig. 15. Predicted and measured earth
pressure at spring line level vs
height of fill.

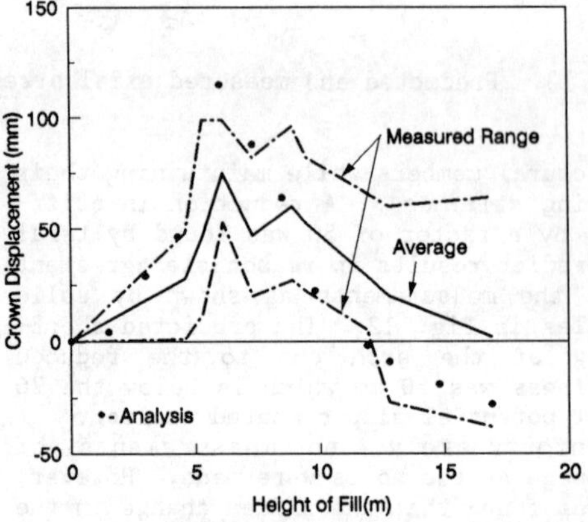

Fig. 16. Measured and predicted culvert
displacements at the crown vs
height of fill.

is occurring.

The predicted and measured displacements of the crown are shown in Fig. 16.
It may be seen that the predictions and
measurements are generally in reasonable
agreement. The crown first moves up as
the side fill is placed, then as backfill
is placed above the crown it moves down
again. It was found that the predicted
displacements just prior to the side fill
reaching the thrust block level were very
sensitive to the soil parameters chosen.
This indicates that great care must be
taken to monitor and control displacements
during the placement of side fill.

The predicted and measured displacements of the spring line are shown in Fig.
17 and are in reasonable agreement. The
measured displacements are inwards at all
stages of loading, while the predicted
displacements are inwards initially during
side filling and later are slightly out-
wards when backfill was placed over the
crown.

The predicted and measured movements of
the crown relative to a point in the soil
at the same elevation of the crown but
offset 8.7 m (Fig. 8) are shown in Fig.
18. Both measurements and predictions
indicated that the crown moved down relative to the soil, although the prediction
slightly overestimates the observed movement. The prediction with the stiff arch
(not shown) indicates that the crown moved
up relative to the backfill, whereas with
reduced stiffness it is predicted to move
down in general agreement with the
measurements.

11 SUMMARY AND CONCLUSIONS

The long-span, high-cover arch culvert at

36

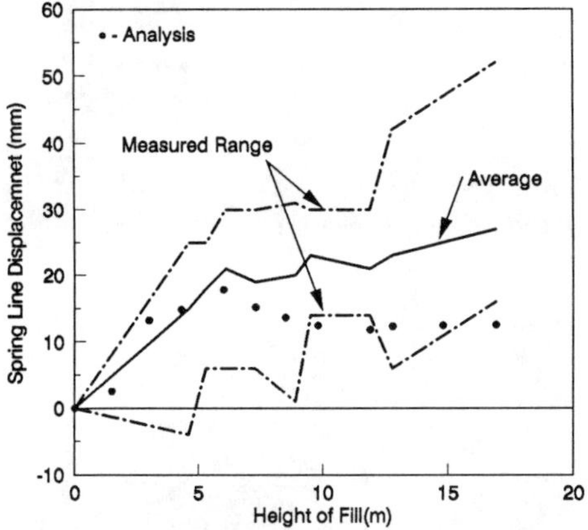

Fig. 17. Measured and predicted displacement of culvert at spring line vs
height of fill.

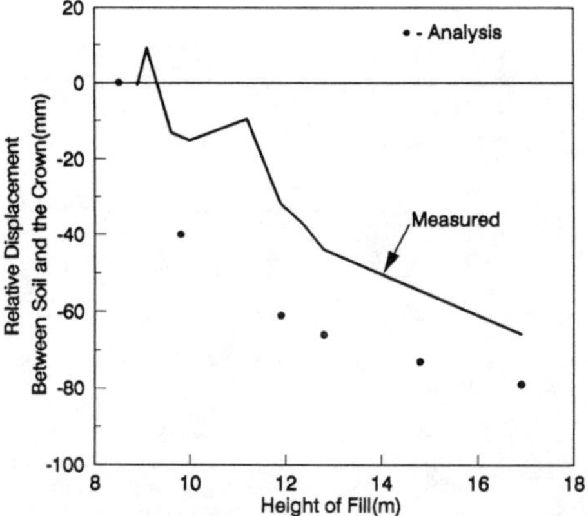

Fig. 18. Predicted and measured relative
displacements between soil and
crown vs height of fill.

Elkhart Creek in British Columbia, Canada,
was instrumented with 3 rings of strain
gauges, load cells, and differential
movement gauges. In addition, movements
of the arch were recorded during and after
construction.

The measurements indicate that the
axial stress in the arch was much lower
than expected from the weight of the
overlying soil. The measured low stresses
in the soil above the crown, and high
stress values outside the arch and
adjacent to the spring line indicate that
positive arching in the soil was occurring
shedding the load from above the arch and
out into the side fills.

Finite element analyses of the soil-
structure system indicate that the low
measured thrust values were caused by slip
at the joints when the fill was about 3 m
above the crown. Such slip resulted in a
greatly reduced stiffness of the metal
arch causing subsequent fill load to be
transferred to the soil in "arching
action", so that little further increase
in axial stress in the metal arch
occurred.

12 ACKNOWLEDGEMENTS

The authors wish to thank Mr. Donald Azar,
ARMTEC, for valuable discussions regarding
the construction of the Elkhart Creek culvert, and Dr. Ian Weir-Jones, for discussions regarding the field measurements.
We acknowledge Mr. F. Salgado's contributions in carrying out analyses to assess
the importance of interface elements on
the analysis. The authors are grateful to
the Ministry of Transportation and
Highways, British Columbia for permission
to publish this paper.

13 REFERENCES

Duncan, J.M., "Behaviour and Design of
Long-Span Metal Culvert Structures",
ASCE Convention, San Francisco,
October, 1977.

LeFebvre, G., Laliberte, L.M. and LeFebve,
L.M., "Measurement of Soil Arching
Above a Large Diameter Flexible
Culvert", Canadian Geotechnical
Journal, Vol. 13, 1975.

Byrne, P.M. and Duncan, J.M., "NLSSIP: A
Computer Program for Nonlinear Analysis
of Soil-Structure Interaction
Problems", University of British
Columbia, Dept. of Civil Engineering,
Soil Mechanics Series No. 41.

Duncan, J.M., Byrne, P.M., Wong, K.S. and
Mabry, P., "Strength, Stress-Strain and
Bulk Modulus Parameters for Finite
Element Analyses of Stresses and Move-
ments in Soil Masses", College of
Engineering, Univ. of California,
Berkeley, Report No. UCB/GT/80-01.

Duncan, J.M. and Chang, C-Y., "Nonlinear
Analysis of Stress and Strain in
Soils", Journal, Soil Mechanics and
Foundations Division, ASCE, Vol. 96,
No. SM5, Proc. Paper 7513, September
1970.

Byrne, P.M., Cheung, H. and Yan. L., "Soil
Parameters for Deformation Analysis of
Sand Masses", Canadian Geotechnical J.,
Vol. 24, No. 3, 1987, pp. 366-376.

Structural Performance of Flexible Pipes, Sargand, Mitchell & Hurd (eds) © 1990 Balkema, Rotterdam. ISBN 90 6191 165 6

A method for evaluating and projecting future movements of in-situ pipes

David C.Cowherd & Vlad G.Perlea
Bowser-Morner Associates, Inc., Dayton, Ohio, USA

ABSTRACT: A method for evaluating the projected future movement of flexible structures is presented. Flexible structures depend on maintaining their shape without becoming too flat in order to maintain their integrity. If a flexible metal structure becomes too flat on the top or sides, it will experience distress and ultimately, collapse. A method for evaluating pipes based on degree of flatness is presented. A technique for determining how flat a pipe can become before remedial action is required has been developed and is discussed. A method for the determination of the degree of flatness based on measurements of chords and mid-ordinates is described. A computer program "MULTSPAN" has been developed to evaluate the degree of flatness and make recommendations as to what should be done in the case of a deflected pipe. The recommendations vary from no action, to lowering the load rating of the road, to closing the road until further action is taken.

A method for evaluating the predicted future movement of a flexible structure is also presented. This method utilizes soil type to make projections of future movement based on conventional foundation settlement theory. Built into the program are average characteristics of seven soil types. It is possible to estimate the surrounding soil properties using only standard penetration data or other information on the degree of compaction. The method utilizes the width of the select backfill, the characteristics of the material beyond the backfill and the characteristics of the structure, to calculate the future movement and over what period of time the movement is expected to occur. It is possible to evaluate a pipe which is deflected and predict future deflection and whether or not the ultimate deflection is likely to create collapse. The equations in this method were calibrated to actual data obtained on several deflected or collapsed structures. A computer program "SOILEVAL" was developed to make the necessary computations manageable.

THEORY OF FLEXIBLE METAL STRUCTURES

Large diameter flexible metal conduits are used for a variety of purposes including underpasses for vehicular traffic, passage of storm water, silos for weapons, underground housing, and many other applications. Since these structures are thin-walled with spans up to 45 feet, designing an installation to safely carry both live and dead loads requires careful evaluation and control of the backfill. Many of these structures have been installed without careful control of the supporting backfill material. As a result, failures have occurred with attendant loss of property and life. Some of these failures occurred even after the structure was given a "clean bill of health" during an annual inspection program.

All flexible buried structures depend on the backfill and the foundation bedding for support. As load is applied to a structure, either from soil loading above the top of the structure or a live load, the structure tries to deflect downward from the top and outward on the sides. As long as the metal is not over-stressed, the sides of the structure must move outward before the structure can significantly deflect downward. In actual fact, since the structure is flexible, some downward movement obviously can occur without the sides of the structure moving outward. However, for major movement to occur without structure failure, the sides of the structure must move outward. As the sides of the structure try to move outward, additional stress is applied to the backfill. Depending on the degree of compaction of the backfill and the bedding, the structure will move to greater or lesser amounts. As long as the backfill is well compacted, the sides or haunches of the structure cannot move sufficiently to allow the structure crown to flatten. In such an instance, there is no danger of structure collapse.

There are basically two modes in which a failure of flexible buried structures can occur:

    1) failure in ring compression, and
    2) failure due to excessive deflection.

These two modes of failure are very different. Ring compression failure is caused by over-stressing and failure of the metal in the structure. Deflection failure is caused by excessive movement of the structure into the backfill allowing the crown of the structure to become flattened causing collapse. Each of these modes of failure is discussed below.

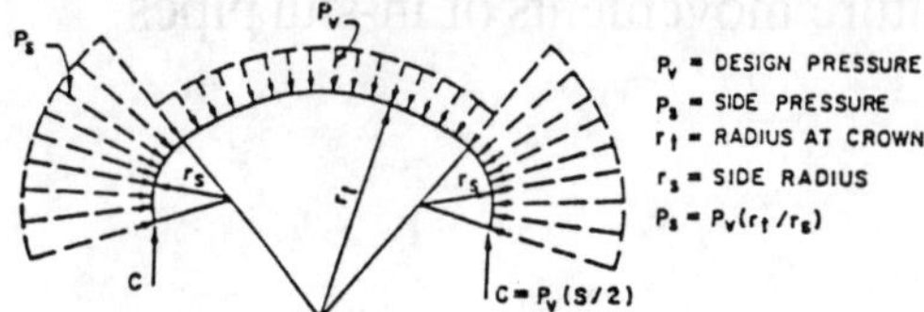

Figure 1 Ring Compression Formula

Ring Compression

The assessment of ring compression is relatively simple and can, for the most part, be made from a review of the plans. Ring compression stress depends only on the external load applied to the structure. The method for calculating ring compression, C, and the associated factor of safety is shown schematically in Figure 1 and discussed in the following paragraphs.

The vertical load (P_v) is calculated from the soil dead load on the top of the pipe plus the live load. The soil dead load is equal to the unit weight of the soil times the depth of cover. The live load can be calculated for any loading configuration. The effects of the live load are reduced with depth.

Knowing the ring compression and the cross-sectional area of the pipe wall, the actual ring compressive stress can be calculated:

$$f_c = \frac{C}{A} \qquad (1)$$

Where:

f_c = actual ring compressive stress (lb/in^2)
C = ring compression (lb/ft)
A = cross-sectional area of plate (in^2/ft)

The ultimate compressive stress, f_b, is determined from the relationship:

$$f_b = f\,(S/r) \qquad (2)$$

Where:

S = span in inches
r = radius of gyration = $\sqrt{I/A}$
I = moment of inertia of the pipe wall (in^4/ft)
A = area of the pipe wall (in^2/ft)

If the S/R ratio is <294, an allowable stress (f_b) of 33,000 psi is utilized. If S/r is in excess of 294, then f_b must be calculated utilizing one of the following formulas:

f_b – 40,000 - 0.081 (S/r)^2, when S/r is between 294 and 500 and

$$fb = \frac{4.93 \times 10^9}{(S/r)^2} ,\ \text{when S/r is >500}$$

The factor of safety is then:

$$\text{F.S.} = \frac{f_b}{f_c} \qquad (3)$$

In the majority of cases, the typical loading encountered by corrugated metal structures in the field results in factors of safety of 6 to 8. This is because the minimum thickness required for constructibility exceeds the long term load requirements. Since this theory only considers loading on the structure and ignores the bending moment stresses resulting from deflections and/or flattening (which is the dominant mode of failure), it should be considered secondary to the primary analysis of deflection stability.

Deflection Stability

As the sides of a pipe deflect, the top of the pipe moves toward flatness. As the pipe approaches the point of being flat, it no longer acts as an arch and the potential mode of failure is no longer ring compression. The degree of flatness can be measured in several ways. It can be determined by measuring the change in mid-ordinate height, the change in radii, or change in the angle of the plates. The easiest measurement to establish is the mid-ordinate height. This is a direct measurement and does not require calculating the radii. In order to totally assess the shape stability, the following measurements are required.

A = Span = Largest Horizontal Dimension of Structure
B = Rise = Height of the Structure Crown Above the Footing or Bottom of the Structure
C = Top Center Chord
D = Top Left Chord
E = Top Right Chord
F = Left Side Chord
G = Right Side Chord
H = Left Corner (or Haunch) Chord*
I = Right Corner (or Haunch) Chord*
J = Bottom Chord Length *
K = Top Center Mid-Ordinate
L = Top Left Mid-Ordinate
M = Top Right Mid-Ordinate
N = Left Side Mid-Ordinate
O = Right Side Mid-Ordinate
P = Left Corner (or Haunch) Mid-Ordinate*
Q = Right Corner (or Haunch) Mid-Ordinate*
R = Bottom Mid-Ordinate*
A1, A2, or C1, C2, or J1, J2 = Approximately One-Half of the Span or Chord Dimensions Defined as A, C, and J above

*These dimensions are not included for an arch resting on footings.

See Figure 2 for the location of these measurements for a typical pipe arch structure.

In order to evaluate the shape stability of a structure, it is necessary to define how flat a structure crown can become before failure will occur. The maximum amount of flattening which can safely

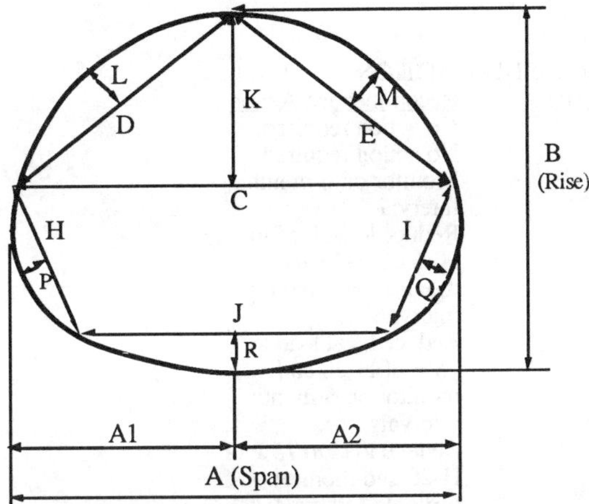

Figure 2 Typical Measurements for a Pipe-Arch Structure.

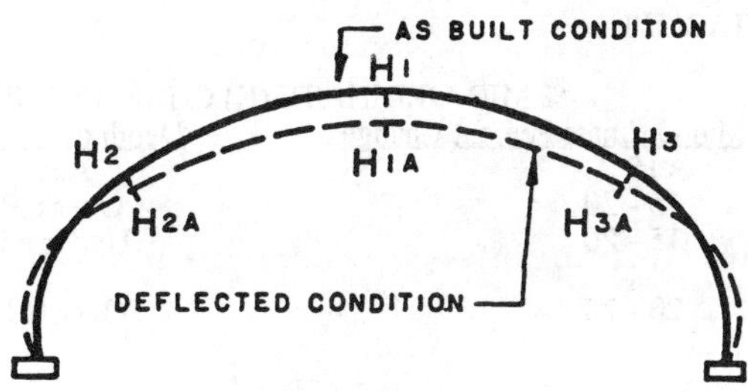

Figure 3 Percent Change in K (Based on Elevation)

occur is a function of many things including the following:

1) shape or type of structure,
2) height of fill over structure,
3) amount of live load,
4) corrugation geometry,
5) gauge or thickness of the structural plate,
6) span,
7) physical condition; and
8) backfill characteristics.

By making a few assumptions, reasonable limits of how flat a structure can become before problems are experienced can be made. These assumptions made in this analysis are:

1) a structure can be considered as a series of curvilinear beams fixed at the points of radius change,
2) the radial load applied around the structure is a function of the ratio of the top radius to the radius under consideration x vertical load (see Figure 1); and
3) the load is applied perpendicular to the structure in all locations (see Figure 1).

By varying the degree of curvature of the top arch, it is possible to reach some degree of curvature at which the structure would fail under the given loading; thus, it is possible to define the degree of flattening that would cause failure under a given set of conditions. While this approach is simplistic it is conservative and it provides a means of evaluating the question of how flat is too flat. The degree of flattening which can be tolerated is dependent on specific site conditions; i.e., the density of the backfill, the thickness of the metal, and the height of fill or loading condition on the structure. Therefore, how flat a structure can become without danger is highly site specific. It is desirable however, to define flatness in some easily quantifiable term that is

recognizable in all cases and to set limits on the amount of flatness that can be tolerated before further action is required. An easily measured parameter which directly indicates the amount of flatness is the mid-ordinate dimension of an arc. The percent change in the mid-ordinate dimension provides a measure of the degree of flatness. Based on calculations as outlined above and field measurements of approximately 100 corrugated metal structures and the observation of several failures, the remedial action required as a function of percent change in mid-ordinate height has been determined (see Table 1).

It may be desirable to determine elevations of the top radius breaks rather than measure mid-ordinate heights. If the measurements are taken using elevations the percent change in the mid-ordinate dimension (K) would then be calculated by the following relationship (see Figure 3).

When using this method of evaluating flatness of the structure, the as-built dimensions should be compared with the plan dimensions to ensure that they are within design tolerances.

It should be noted that it is possible for there to be a significant change in the elevation difference of Hl and HlA and yet only a small percent change in the mid-ordinate dimension K. In such a movement, the entire crown of the structure would have shifted downward while maintaining its original top curvature. In order for such a movement to occur, the sides of the structure would need to move outward as shown in Figure 3. Such a movement would not necessarily represent a threat to the structural integrity of the span since the original crown curvature has been maintained.

While flattening of the crown is of primary concern when evaluating the structural integrity of corrugated metal structures, there are additional deflection conditions which could cause problems; they are: leaning of the structure and flattening of the sides.

MULTSPAN

A computer program (MULTSPAN) has been developed to evaluate the degree of flattening. The program analysis provides recommendations for remedial action based on the degree of flatness of top arcs within the structure. Most of these structures

TABLE 1

% MID-ORDINATE (K) CHANGE AND REMEDIAL ACTION

Mid-Ordinate Percent Change	Depth of Cover (ft)	Recommended Action
<15	Any	No action required.
15 - 20	Over 6.0	No action required.
15 - 20	Under 6.0	Monitor on 6-month interval.
20 - 25	Over 6.0	Reduce legal load to 90% of H-20 and monitor on 6-month intervals.
20 - 25	Under 6.0	Reduce legal load to 75% of H-20 and monitor on 6-month intervals.
25 - 30	Over 6.0	Reduce load to 75% of H-20 and monitor on 6-month intervals.
25 - 30	3.0 - 6.0	Reduce load to 50% of H-20 and monitor on 6-month intervals.
25 - 30	Under 3.0	Reduce load to 50% of H-20 and do detailed analysis.
>30	Any	Close road until detailed analysis is done.

are various portions of circles. It is reasonably simple to measure chords and mid-ordinates and then compare these chords and mid-ordinates to the design value. Design values for various types of structures are built into the program or other values can be input. The program compares the actual measured values to the design values (or can compare them to the measured values from previous inspection) and calculates the deflection as a percentage of mid-ordinate change. Figure 2 shows a typical arch structure and the measurements which are made with such a structure.

The measurements are entered into the program and the degree of flatness defined as a percentage change of the mid-ordinate is computed. Table 1 outlines the recommendations for various degrees of top mid-ordinates.

Typically, the mid-ordinates of the top of the pipe are considered in the recommendations and not those on the sides. It is possible, however, to output all mid-ordinate changes and compare them to design values. Table 1 was developed on the basis of thin shell theory and on empirical data. Some 100+ pipes were evaluated and used in the evaluation of Table 1. Some six collapsed structures were also used in this evaluation to determine how flat a structure can be before problems develop. By modeling the programs on actual failed and deflected structures, the soil-structure interaction is more effectively taken into account.

By using the method, it is possible to evaluate a structure by making measurements of chords and mid-ordinates, compare them to design values, and obtain an evaluation of whether or not remedial action is required. The program also calculates the factor of safety in relationship to ring compression. Such things as rusting out of a portion of the pipe or cracking around bolts can be taken into account in this evaluation. The program provides recommendations as to what should be done relative to the structure.

SOILEVAL

A method was developed to evaluate the potential for continued movement of a structure and predict a general time frame for the movement. The model is based on classical soil mechanics theory, and compared to actual field cases for calibration. An attempt was made to use simplified data for input to allow the program to be simple and useful by field personnel. It is possible to use the program to either evaluate an existing structure or as a design tool to evaluate the type, required width, and degree of compaction of the select backfill and the original soil. A computer program "SOILEVAL" was developed to make the necessary computations manageable

Background

The deflection of a buried flexible structure can be predicted in accordance with the classical formula for evaluation of strain or deformation of a structural member (strain = stress/modulus of elasticity) *(1)*. The equation for deflection of a flexible structure supported by backfill takes the following form:

$$\frac{\text{Structure}}{\text{Deflection}} = \frac{\text{Load On Structure}}{\text{Structure Stiffness} + \text{Soil Stiffness}} \qquad (5)$$

Several theories to evaluate the structure-soil interaction have been proposed. Most theories for structure/soil interaction use some form of this equation.

Iowa Formula

The formula recommended by Spangler and Watkins *(2)* for predicting deflections of buried flexible pipe is:

$$\Delta x = \frac{D_l \ K \ W_c \ r^3}{EI + 0.061 \ E'r^3} \qquad (6)$$

Where:

Δx = horizontal deflection of the pipe, considered the same as the vertical deflection (in)

D_l = deflection lag factor (normally taken as 1)

K = a bedding constant, with values between 0.08 and 0.11 , depending on the bedding condition

W_c = vertical load per unit length acting on the top of the pipe (lb/in.)

r = mean radius of the pipe (in)

E = modulus of elasticity of the pipe material (lb/in^2)

I = moment of inertia per unit length of cross section of the pipe wall (in^4/in)

E' = modulus of soil reaction (lb/in^2)

Rearranging Eq. 6, the following formula is obtained:

$$\Delta x = \frac{D_l \ K \ W_c/E'}{PS_r + 0.061} \qquad (7)$$

Where:

PS_r = $EI/(E'r^3)$ = pipe - soil stiffness ratio *(3)*

The use of this formula is currently recommended for the design of corrugated metal pipes *(4)*, smooth steel pipes *(5)* polyethylene pipes *(6)*, PVC pipes *(7)*, fiberglass pipes *(8)*, and generally for all buried flexible structures. Although it is confirmed that E' is not a constant but varies with depth of installation *(9)* the original suggested values by Howard *(1)* are extensively used.

Other Proposed Models

Watkins et al *(6)* summarized a number of proposed variations of the classical Iowa Formula and found that all of them can be represented by a simple relationship:

$$\Delta y = \frac{\varepsilon_s \ (2r)}{A + BEI/(E_s r^3)} \qquad (8)$$

Where:

ε_s = P/E_s = vertical soil strain (soil deformation)

P = $W_c/(2r)$ = vertical nominal pressure acting on top of pipe (lb/in^2)

E_s = soil stiffness (lb/in^2)

A and B = empirical constants

From field tests it was found that much of the deformation process takes place as the backfill is being placed, so for pipe deflection calculations a short-term modulus should be used. However, it is the authors experience that there may be continued deformation with time due to consolidation of the sidefill (especially with cohesive backfills), so that usually ultimate soil settlement should be included in ε_s *(6)*.

More recently several procedures for flexible pipe deformation evaluation have been developed using finite element methods. CANDE (Culvert Analysis, Design) is a Federal Highway Administration sponsored computer program by Katona et al *(10)*. Three levels of sophistication are available: Level 1 is based on closed-form elasticity solution, whereas Levels 2 and 3 are based on the finite element method. SCI (Soil/Culvert Interaction) Design Method by Duncan *(11)* utilizes design graphs and formulas based on finite element analysis. Seed and Duncan *(12)* in the program SSCOMP utilize nonlinear finite element analysis to model soil-structure interaction, taking into account the layerwise placement of backfill and the non-linear stress-strain soil behavior.

The finite-element evaluations are beyond most field personnel and a simplified model for a rough evaluation of flexible buried structures which could be easily used in routine evaluation of existing structures and preliminary design of new structures was desirable. The model was developed to provide this simplified model. The model uses the classical theories of soil mechanics. Some coefficients were however empirically obtained by statistical processing of actual field measurements.

SOILEVAL Model

The form of the equation used in the SOILEVAL model to evaluate soil-structure interaction is:

$$\Delta y = \frac{A \ \Delta W \ SF}{I/r_a^3 + B(E'/E)} \leq F_s \Delta W \ SF \qquad (9)$$

Where:

Δy = maximum expected deflection at the crown of the structure (in.)

F_s = factor of safety to take into account the variability of soil properties equal to 1.5

ΔW = potential horizontal movement of one side of the structure, due to compression of both backfill and original soil under the stress generated by the pipe (in.)

SF = shape factor, defined as the ratio between the vertical displacement of the structure at the crown and the corresponding maximum movement on

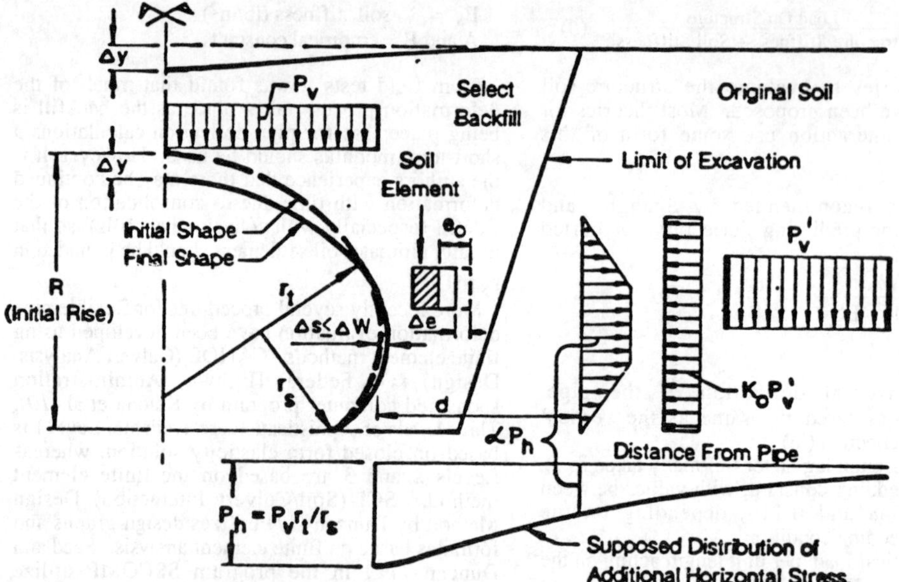

Figure 4. Pipe - Soil Interaction

one side of the structure

r_a = average radius of the structure, equal to the span plus the rise divided by 4 (in.)

A and B = empirical coefficients, statistically determined from field measurements.

All the other parameters are the same as defined before.

The model incorporates the following characteristics of soil/structure interaction:

1. The compressibility index, coefficient of pressure at rest, initial void ratio, relative density or degree of compaction of the soil supporting the structure are used in calculations, instead of special defined characteristics (deformation lag factor, modulus of soil reaction).

2. Both the backfill and the original soil are considered to interact with the structure, up to a distance of about 2.5 times the rise of the structure laterally from the pipe wall.

3. The pipe stiffness is taken into account when it potentially provides resistance to deformation. However, for very flexible pipes the entire potential deformation of the surrounding soil multiplied by a factor of safety is used to determine the maximum expected pipe deflection.

4. A shape factor which represents the ratio between the decrease in rise and the corresponding lateral displacement on one side of the pipe is defined; this shape factor is highly dependent on the shape and dimensions of the pipe structure.

How each parameter is determined in the model is described as follows.

Potential Horizontal Movement Due to Soil Compressibility (ΔW)

The classical theory of settlement for shallow foundations has been used for calculating the potential horizontal movement. The fill at the side of the structure is considered as a soil column, loaded by the pressure created by the structure onto the fill *(13)*. Figure 4 explains the meaning of the notation used.

The potential horizontal displacement of the structure due to soil compressibility (without any restriction due to structure stiffness) is obtained by totaling the displacements calculated for each incremental layer on one side of the pipe. Similar to the calculation of shallow foundation settlement, the summation is extended to a distance where the additional stress in the soil generated by the pipe is less than 20% of the horizontal stress corresponding to the overburden pressure, or to a maximum distance of 2.5 times the rise dimension, whichever is less. The following equations are used:

$$\Delta W = \Sigma \left[W \, \Delta e / (1 + e_0) \right] \qquad (10)$$

$$\Delta e = C_c \, \log \frac{K_0 P'_v + \alpha \, P_h}{K_0 P'_v} \qquad (11)$$

$$\alpha = 10^{(-0.45 \, d / R)} \qquad (12)$$

$$P_h = P_v \, r_t / r_s \qquad (13)$$

Where:

W = initial width of an incremental layer (in.)

Δe = potential decrease in void ratio

e_0 = initial void ratio, not affected by the supplementary pressure induced by the structure

44

Table 2

COMPRESSIBILITY INDEX VALUES

CATEGORY OF SOIL	TYPE OF SOIL	ASTM D-2487 CLASS	FINES CONTENT (% < #200 SIEVE)	C_c Values For:		
				LOOSE/SOFT MATERIAL ($C_{c,w}$)	MEDIUM ($C_{c,av}$)	DENSE/STIFF MATERIAL ($C_{c,b}$)
I	Gravel	GW,GP	<12	0.03	0.01	0.003
II	Silty/Clayey Gravel	GM,GC	12 – 50	0.05	0.02	0.008
III	Well-Graded Sand	SW	<12	0.06	0.02	0.007
IV	Poorly-Graded Sand	SP	<12	0.05	0.03	0.018
V	Silty/Clayey Sand	SM,SC	12 – 20	0.06	0.03	0.015
			20.1– 30*	0.16	0.08	0.040
			30.1– 50	0.33	0.17	0.088
VI *** (min 0.025)	Silty Soils	ML,MH	>50	** (max 0.40)	$0.007 (W_L{-}10)$ but min 0.05 and max 0.20	
VII *** (min 0.050)	Clayey Soils	CL,CH	>50	** (max 0.80)	$0.007 (W_L{-}10)$ but min 0.10 and max 0.40	

NOTES: W_L is the liquid limit of the soil
    * Default value is 30 when grain size distribution is not given
    ** $C_{c,w} = 2\ C_{c,av}$
    *** $C_{c,b} = (C_{c,av})^2 / C_{c,w} = 0.5\ C_{c,av}$
    If W_L is unknown, default values are used, which give:
    $C_{c,av} = 0.10$ for Category VI soil
    $C_{c,av} = 0.18$ for Category VII-a soil (lean clay)
    $C_{c,av} = 0.35$ for Category VII-b soil (fat clay)

C_c = compression index of the soil (backfill or original soil beyond the backfill)

K_o = the coefficient of earth pressure at rest

P_v' = the effective overburden pressure at the level of calculation; i.e., approximately in the middle of the loaded area by the structure (lb/in^2)

α = the influence coefficient at the distance d(in) from the structure corresponding to the middle of a given incremental layer; it is based on a procedure previously developed by Cowherd (*14*) for round pipes, using the rise instead of pipe diameter.

P_h = the supplementary pressure on the side plates of the structure induced by the downward movement of the structure's crown (lb/in^2)

P_v = the total vertical pressure due to the soil dead load on the top of the structure, approximately considered equal to the unit weight of the backfill times the depth of cover (lb/in^2)

R = rise of the structure (in)

r_t = top radius of the structure(in)

r_s = Side radius of the structure (in)

A number of soil properties must be known or properly estimated, both for the backfill or the original soil. They are: e_0, C_c, K_o, and the unit weight. The program gives the option of entering these as input data or to estimate them based on various levels of knowledge of soil condition.

Potential backfill and original soils have been divided into seven categories and the geotechnical parameters estimated for each category. These parameters versus soil type have been built into the program so that by choosing a soil category based on simplified soils data, the appropriate geotechnical indexes are automatically used. If more precise data are available, the program allows the direct input of the soil parameters. Table 2 shows the soil categories and corresponding C_c values built into the model.

The relative degree of compaction of the backfill (stiffness or hardness of original soil) is necessary for an evaluation of the potential movement of a structure. The method provides three approaches to estimate the average degree of compaction (denseness/hardness) of soils; i.e.:

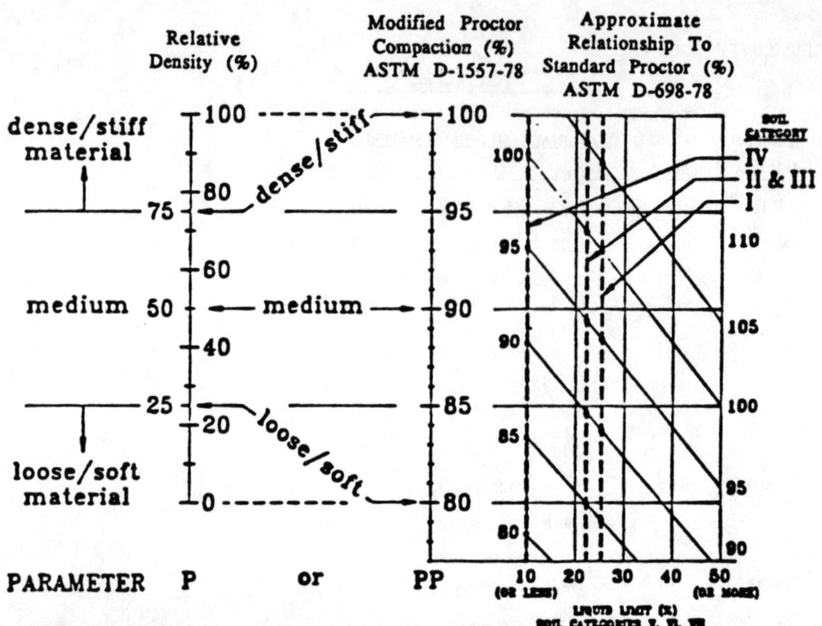

Figure 5. Approximate Comparison Between Various Estimates of Denseness or Stiffness

Table 3

RELATIVE DENSITY/CONSISTENCY PARAMETER (P) FOR VARIOUS SOIL TYPES

SOIL CATEGORY	AVERAGE D_{50} (mm)	P: If P > 75, use P = 75 If P < 25, use P = 25
I	12.5	$43 \times \log [96.56 \times N \times D_{50}^{-0.284} \times (\sigma v')^{-0.56}]$
II	1.0	$43 \times \log [72.42 \times N \times D_{50}^{-0.284} \times (\sigma v')^{-0.56}]$
III	0.5	$43 \times \log [65.19 \times N \times D_{50}^{-0.284} \times (\sigma v')^{-0.56}]$
IV	0.3	$43 \times \log [60.36 \times N \times D_{50}^{-0.284} \times (\sigma v')^{-0.56}]$
V	0.15	$43 \times \log [54.32 \times N \times D_{50}^{-0.284} \times (\sigma v')^{-0.56}]$
VI	0.023	$43 \times \log [36.21 \times N \times D_{50}^{-0.284} \times (\sigma v')^{-0.56}]$
VII	0.007	$43 \times \log [24.14 \times N \times D_{50}^{-0.284} \times (\sigma v')^{-0.56}]$

- standard penetration test results;
- direct measurements of soil density and moisture content; and
- design requirements or inspection records which give the degree of compaction

A parameter (P or PP)which is a measure of relative density, degree of compaction or stiffness, has been developed to allow a selection of soil parameters based on the stiffness of the backfill. Figure 5 shows the correlation between various possible criteria for defining the denseness or the stiffness of the soil.

When the standard penetration test is used to characterize the soil condition, the equations in Table 3 are used to estimate the parameter P.

Where:

P = a parameter which is used in the same manner as the relative density for estimation of the compressibility index.

N = blows per foot in Standard Penetration Test

D_{50} = mean diameter of soil particles (mm)

$\sigma v'$ = effective overburden pressure at the test location (psf)

The formulas in Table 3 have been based on relationships suggested in literature *(14, 15, 16, 17)*. The equations shown in Table 3 have been built into the model to calculate "P" when the D_{50} and average standard penetration value "N" of the backfill are known.

46

Table 4

INITIAL VOID RATIO FOR VARIOUS SOIL TYPES

SOIL CATEGORY	TYPE OF SOIL	ASTM D-2487 CLASS	STANDARD PENETRATION BLOW COUNT, N	VOID RATIO, e_O	K_O FOR: BACKFILL	K_O FOR: ORIGINAL SOIL
I & II	Gravels	GW,GP	≤ 10	0.6	0.4	0.4
		GM,GC	11-30	0.5	0.5	0.45
			≥ 31	0.4	0.6	0.5
III & IV	Sands	SW,SP	≤ 10	0.7	0.4	0.4
			11-30	0.55	0.5	0.45
			≥ 31	0.4	0.6	0.5
V	Silty/Clayey Sand	SM,SC	≤ 10	0.8		
			11-30	0.6	0.6	0.5
			≥ 31	0.45		
VI	Silty Soils	ML,MH	≤ 5	0.9		
			6-15	0.7	0.6	0.5
			≥ 16	0.5		
VII-a	Clayey Soils, $W_L < 50$	CL	≤ 5	1.0		
			6-10	0.8	0.7	0.6
			≥ 11	0.6		
VII-a	Clayey Soils, $W_L \geq 50$	CH	≤ 5	1.6		
			6-10	1.1	0.7	0.6
			≥ 11	0.7		

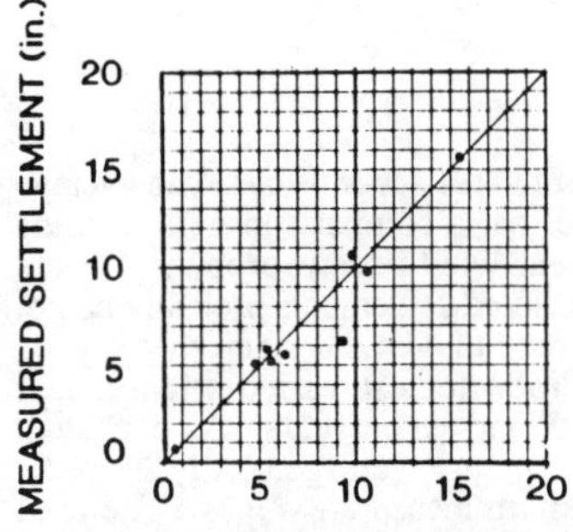

Figure 6 Model Calibration:

The parameters e_O and K_O may be directly input into the model, if known. If these values are not known, the model provides a value based on soil category and standard penetration values (or other parameters that give an estimate of the degree of compaction) as shown in Table 4.

Shape Factor

The shape factor establishes a relationship between movement of the crown and movement of the sides of the structure. It is defined as the ratio between the decrease in rise (Δy) and the corresponding increase in half-span (or in half-chord at the mid-height of the rise in the case of circular arches, where the span is measured at the foundation level). Simple geometrical relationships have been derived considering small, but finite deformations of the pipe when chord length may be considered a constant.

These relationships are listed in Table 5. An empirical factor equal to 0.84 was derived to correct the theoretical formula for small pipe-arches, based on over 300 field measurements.

Empirical Constants A and B

The empirical parameters A and B in Equation g were statistically determined from field measurements. Ten very carefully monitored case histories were used to estimate these values. The best fit between the measured settlements at the crown and the computed values was found for A = 1.2×10^{-7} and B = 1.5×10^{-3}. These values gave the minimum scatter of data, as shown in Figure 6 where measured settlements at crown is compared with the estimated decrease in rise.

Use of SOILEVAL Model with Buried Structures Other than Steel

Any empirical formula should be applied to cases where conditions are similar to those of the case histories used in the determination of the parameters. However, the model like the Iowa Formula, has a theoretical substantiation and the result depends on such characteristics as elastic moduli of the materials which interact. The Iowa Formula was derived for steel pipes of small diameters, but subsequently was successfully used in corrugated steel or aluminum superspan structure design as well as in connection with pipes of other materials. For example, the Iowa Formula is a standardized method for the design of fiberglass pipes *(8)* and is recommended for PVC pipes analysis also *(7)*. Therefore, the validity of the model was investigated for conditions which

Table 5

A SUMMARY OF SHAPE FACTORS FOR VARIOUS TYPES OF PIPES

PIPE SHAPE	CHARACTERISTICS	SHAPE FACTOR	RANGE, (AVERAGE VALUE) FOR STANDARD PIPES	
Roundpipe	$S = R$	2.0		(2.0)
Longspan Pipearch	$S \geq 20$ Feet	$0.5 \left(\dfrac{S}{R-R_T} + \dfrac{S}{R_T} \right)$	2.97–4.90	(4.0)
Small Pipe-Arch	$S < 20$ Feet	$0.5 \left(\dfrac{0.84\,S}{R-R_T} + \dfrac{S}{R_T} \right)$	2.53–3.99	(3.3)
Underpass	$R_T > R_B$	$0.5 \left(\dfrac{S}{R_B} + \dfrac{S}{R_T} \right)$	2.28–2.60	(2.4)
Structural Plate Arch	$R < \dfrac{S_B}{2}$	$\sqrt{2\left(\dfrac{S_B}{R}\right)^2 + 4}\; - \dfrac{S_B}{R}$	1.42-2.72	(2.0)
Vertical Ellipse	$R > S$	$\dfrac{2S}{R}$	1.79-1.82	(1.8)
Horizontal Ellipse	$S > R$	$\dfrac{2S}{R}$	2.70-3.82	(3.1)
Low Profile Arch	$R_T \cong 0.9R$	$0.5 \left(\dfrac{S - S_B}{R - R_T} + \dfrac{S}{R_T} \right)$	1.46-1.83	(1.7)
High Profile Arch	$R_T \cong 0.6R$	$0.5 \left(\dfrac{S - S_B}{R - R_T} + \dfrac{S}{R_T} \right)$	1.61–1.95	(1.9)
Pear	$R_B > R_T$	$0.5 \left(\dfrac{S}{R_B} + \dfrac{S}{R - R_B} \right)$	1.88–2.30	(2.1)

Notations in Table 5:

S	=	Span	R = Rise	
S_B	=	Bottom Span	R_T = Top Rise	
R_B	=	$R - R_T$ = Bottom Rise	r_t = Top Radius	

substantially differ from large span structures of corrugated steel as used in the calibration. (It is noted that a preliminary formula instead of the final equation 9 was used in computations.)

The CONTECH book on "ARMCO Truss Pipes" *(18)* makes available the results of deflection measurements on about 140 thermoplastic composite pipes used for gravity-flow sanitary sewer systems. These pipes consist of a double-wall system, with concentric inner and outer walls braced by a truss-type structure. The walls and the truss structure are formed of a single thermoplastic extrusion of either polyvinyl chloride (PVC) or acrylonitrile butadiene styrene (ABS). The composite pipe stiffness defined by the ration $EI/0.149r^3$ is about 200 psi.

Little information was available about the backfill soil condition, the trench dimensions and the compaction quality. Since there was no information about the original soil, it was considered indeformable. A rather large width of the trench of 1.5 times the pipe diameter on each side of the pipe was assumed along with an average influence coefficient of pressure distribution in the backfill of 0.5. An approximate correlation between the four ARMCO defined soil classes and the seven categories used in SOILEVAL was assumed. Other approximations were also necessary in defining the denseness and the moisture content of the material used as backfill.

Figure 7 shows the comparison between average and maximum values of the measured settlement and the results of the calculations.

The estimation of the deflection is based on average soil conditions, so that a comparison with average measured settlement would seem proper. On the other hand, the measured settlements probably do not represent the ultimate movement since the elapsed time from installation to test varied widely; i.e., between one month and seven years. As expected, the estimated settlements were generally between average and maximum measured values. The total average estimated value for all 136 pipes is 0.013 feet which is between the measured values of 0.007 feet (average) and 0.016 feet (maximum). With such small values of settlement and many assumed conditions the agreement between calculated and measured data is considered satisfactory.

Other application of the model to conditions different from the calibration ones refers to 28 case histories of high density polyethylene pipes reported by Chua and Petroff*(19)*. As shown in Figure 8, in almost all cases the predicted maximum deflection after complete consolidation of the soil was slightly greater than the maximum observed deflections, which also supports the validity of the model in these conditions *(20)*.

Use of SOILEVAL Model for Design of Backfill

The method can be used to evaluate the effect of soil type and degree of compaction/stiffness in both select backfill and original soil *(21)*. For this purpose, the behavior of a very flexible ellipse pipe (24.5-foot span, 14-foot rise) constructed of

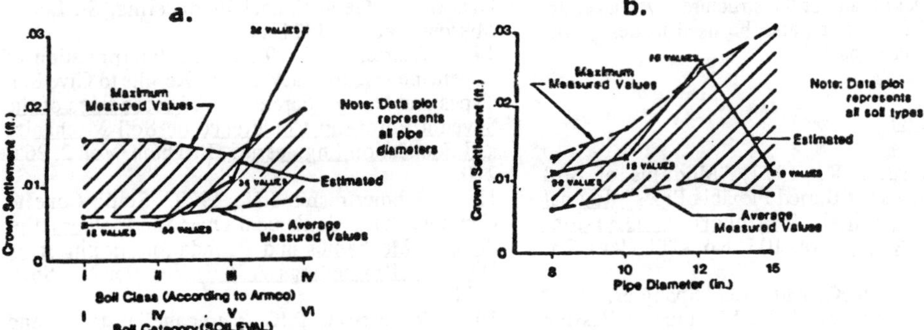

Figure 7 Crown Displacement of Small (<15 inch) Truss Pipe
for Various Soil Types (a) and Pipe Diameters (b)

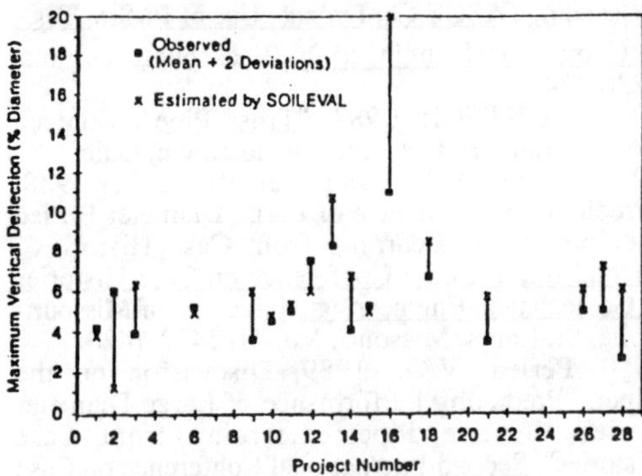

Figure 8 Comparison Between Measured and
Predicted Maximum Deflections of HDPE Pipes

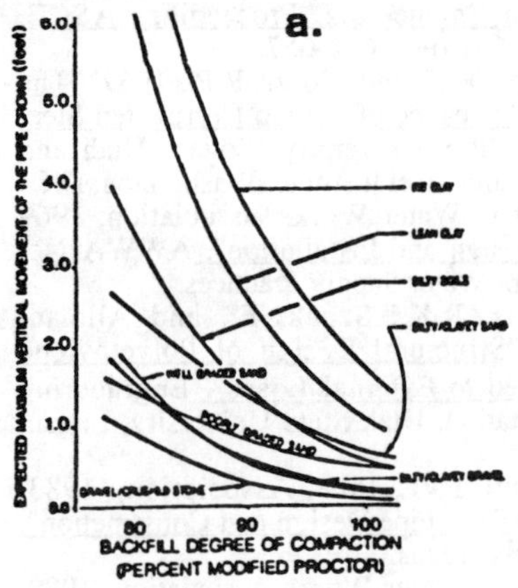

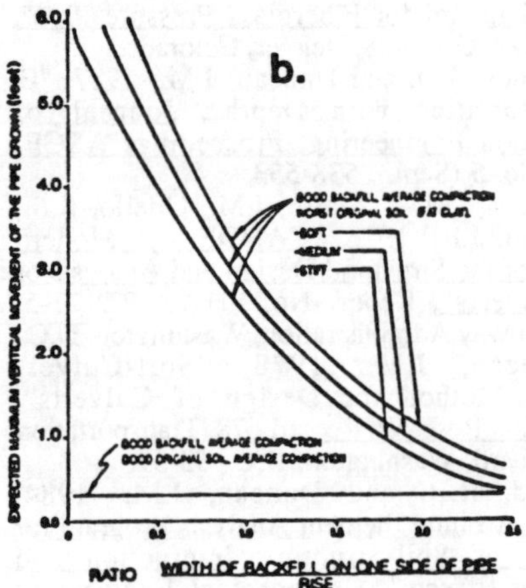

Figure 9 Behavior of a Very Flexible Steel
Structure Under Various Soil Conditions:
  a. Effect of type of backfill of infinite
       width at various degrees of compaction
  b. Effect of select backfill width in various
       original soil conditions.

corrugated steel in various soil conditions was
studied.

The results of this study are summarized in Figure
9. It is evident from this exercise that granular
backfill materials are capable of providing adequate
support even at low degrees of compaction, while
some cohesive materials might not provide adequate
support even at high degrees of compaction.

CONCLUSIONS

1. A method has been developed which evaluates
flexible structures based on shape evaluation by
measurements of chords and midordinates so that
problems can be anticipated long before they occur
and remedial action can be taken.
2. A method has been developed which provides a
simplified semi-empirical method of predicting
movement of flexible buried structures. This method
has been found useful in predicting future
movements of pipes that have experienced some
degree of movement.
3. The soil model can be used as a design tool to

49

design the backfill and/or the structure given a certain type of backfill. It can also be used to design the width of the select backfill.

REFERENCES

1.	Howard, A.K., 1977, "Modulus of Soil Reaction Values for Buried Flexible Pipes", Journal of the Geotechnical Engineering Division, Proceedings ASCE, Vol. 103, No. GT1 (Jan), 33-43.
2.	Watkins, R.K., and M.G. Spangler, 1958, "Some Characteristics of the Modulus of Passive Resistance of Soil: A Study in Similitude", Proc. Highway Res. Board, 37,576.
3.	Jeyapalan, J.K., Ethiyajeevakaruna S.W., and Boldon, B.A., 1987, "Behavior and Design of Buried Very Flexible Plastic Pipes", Journal of Transportation Engineering, Proceedings ASCE, Vol. 113, No. 6 (Nov.), 642-657.
4.	Watkins, R.K. and Moser, R.P., 1969, "The Structural Performance of Buried Corrugated Steel Pipes", Utah State University, Logan, Utah and American Iron and Steel Institute, Washington, D.C.
5.	American Water Works Association, 1969 "Steel Pipe Design and Installation", AWWA No. M11, Manual of Water Supply Practices.
6.	Watkins, R.K., Szpak, E., and Allman, W.B., 1974, "Structural Design of Polyethylene Pipes Subjected to External Loads", Engineering Experiment Station, Utah State University, Logan, Utah.
7.	Uni-Bell PVC Pipe Association, 1983, "Handbook of PVC Pipe Design and Construction", Uni-Bell, Dallas, Texas.
8.	American Water Works Association, 1989, "AWWA Standard for Fiberglass Pressure Pipe", ANSI/AWWA C 950-88, Denver, Colorado.
9.	Hartley, J.D. and Duncan, J.M., 1987, "E' and Its Variation with Depth", Journal of Transportation Engineering, Proceedings ASCE, Vol. 113, No. 5 (Sept.), 538-553.
10.	Katona, M.G., Smith, J.M., Odello, R.S., and Allgood, J.R., 1976, "CANDE - A Modern Approach for the Structural Design and Analysis of Buried Culverts", Report No. FHWA-RD-77-5, Federal Highway Administration, Washington, D.C.
11.	Duncan, J.M., 1978, "Soil-Culvert Interaction Method for Design of Culverts", Transportation Research Record 678, Transportation Research Board, Washington, D.C., 53-59.
12.	Seed, R.B. and Duncan, J.M. 1984, "SSCOMP: A Finite Element Analysis Program for Evaluation of Soil-Structure Interaction and Compaction Effects", Geotechnical Engineering Research Report No. UCB/GT/84-02, University of California, Berkeley, California.
13.	Cowherd, D.C., Thrasher, S.M., Perlea, V.G., and Hurd, J.O., 1988, "Actual and Predicted Behavior of Large Metal Culverts", Proceedings of the Second International Conference on Case Histories in Geotechnical Engineering, St. Louis, Missouri, Vol. 2, 1471-1476.
14.	Searle, J.W., 1979, "The Interpretation of Begemann Friction Jacket Cone Results to Give Soil Types and Design Parameters", Proceedings of the Seventh European Conference on Soil Mechanics and Foundation Engineering, Brighton, Vol. 2, 265-270.
15.	Schmertmann, J.H., 1979, "Static Cone to Compute Static Settlement Over Sand", Journal of the Soil Mechanics and Foundation Engineering Division, Proceedings ASCE, Vol. 96, No. Sm3, 1011-1043.
16.	Robertson, P.K., Campanella, R.G., and Wightman, A., 1983, "SPT-CPT Correlations", Journal of Geotechnical Engineering, Proceedings ASCE, Vol. 109, No. 11, 1449-1459.
17.	Solymar, Z.V., and Reed, D.J., 1986, "Comparison Between In-Situ Test Results", Proceedings ASCE Conference, Use of In-Situ Tests in Geotechnical Engineering, Blacksburg, Virginia, 1236-1248.
18.	CONTECH, 1988, "Truss Pipe", Contech Construction Products, Inc., Middletown, Ohio
19.	Chua, K.M. and Petroff, L.J., 1988, "Predicting Performance of Large-Diameter Buried Flexible Pipes: Learning from Case Histories", Second International Conference on Case Histories in Geotechnical Engineering, University of Missouri-Rolla, St. Louis, Missouri, Vol. II, 1417-1420.
20.	Perlea, V.G., 1989, Discussion on the Paper, "Predicting Performance of Large-Diameter Buried Flexible Pipes: Learning from Case Histories", Second International Conference on Case Histories in Geotechnical Engineering, University of Missouri-Rolla, St. Louis, Missouri, Vol. II, 1417-1420.
21.	Cowherd, D.C. and Perlea, V.G., 1989, "An Evaluation of Flexible Metal Pipes," Ohio County Engineers News, Spring, 10-11 and 16.
22.	Cowherd, D.C., 1972, "Factors to be Considered in the Design of Backfill for Large Diameter Flexible Metal Conduits", Proceedings 1972 Ohio River Valley Soil Seminar "Lateral Earth Pressures - Case History Studies", Ft. Mitchell, Kentucky

Structural Performance of Flexible Pipes, Sargand, Mitchell & Hurd (eds) © 1990 Balkema, Rotterdam. ISBN 90 6191 165 6

Application of deep corrugated steel plate to a new bridge in Tennessee

Samuel C. Musser
Syro Steel Company, Centerville, Utah, USA

Sam B. Graham
Southeast Mat Company, Crossville, Tenn., USA

Terry C. Miller
TARE, Inc., Crossville, Tenn., USA

ABSTRACT: A case study is presented which evaluates the economic viability of corrugated metal long span arches as compared to conventional bridges. A number of site factors are shown to impact the cost/benefit analysis of bridges including road profile, site terrain, environmental regulation, and availability of construction materials. To assure least cost, these factors should be carefully evaluated on a case-by-case basis. Long span arches are shown to be cost effective alternatives to conventional bridge structures in the short to medium span range.

Corrugated metal long spans are soil-structure interaction systems which can be successfully and conservatively modeled using computerized analysis tools. The design should include the specification and field quality control of both the corrugated metal shell and the surrounding soil envelope. Also, some aspects of the AASHTO specifications for long span design are shown to be inadequate for application to large long span bridges.

1 PROJECT BACKGROUND

A new bridge was required as part of a 78 acre industrial/commercial development project owned by William F. Graham. The property to be developed was situated in Cumberland County, approximately two miles north of Crossville, Tennessee. The bridge would cross the Obed River as part of a 0.65 mile extension of Northside Drive. TARE, Incorporated of Crossville was the consulting engineer for the owner.

The new road and bridge connects two major highways serving as direct routes to Crossville, US 127 on the east and US 70N on the west. Figure 1 depicts a vicinity map of the area. The new road complies with the State of Tennessee master plan for highway development in the area and will become part of the State highway system upon completion and approval.

The bridge and road construction project fell under the jurisdiction of five governmental regulatory agencies. The first two levels of approval came from the Cumberland County Planning Department and the County Road Superintendent. The Tennessee Department of Transportation stipulated design volume, loading and geometry considerations for the bridge and road. The US Army Corps of Engineers and the Tennessee Valley Authority dictated the hydro-

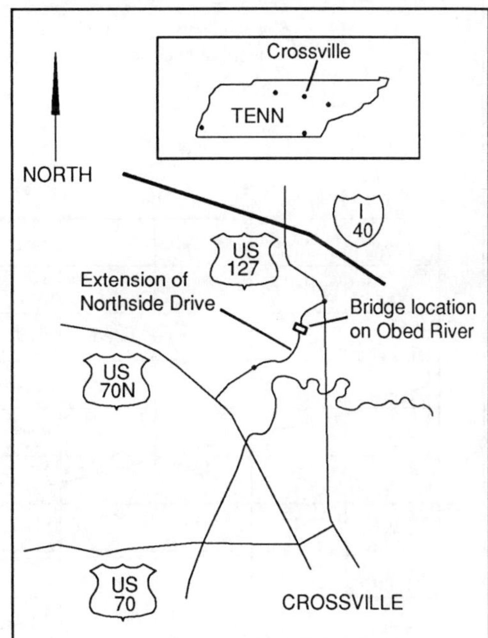

Figure 1. Project vicinity map.

logic and hydraulic design of the channel as well as the manner of construction in and around the Obed River.

2 SITE DESIGN CONSIDERATIONS

The Tennessee DOT master plan anticipated a 2-lane highway for this route with a volume under 400 vehicles per day. The bridge design load was specified as HS20-44 in accordance with AASHTO bridge standards.

The topography of the site is typical of the Cumberland Plateau, with undulating contours of heavily wooded land. Elevation above sea level varies along the road alignment from 1,773 at the tie-in with US 127 to 1,689 at the flowline of the river, a difference of 84-feet. The roadway plan and profile in the vicinity of the bridge are given in Figure 2.

Bedrock occurs at relatively shallow depths, covered by deposits of fined grained silts, clays and organic top soils. Borrow materials for engineered fills must generally be imported as crushed limestone or decomposed sandstone.

The Obed River has formed a channel approximately 45-feet wide and 4-feet deep in the soft soils. Since the bed of the channel is primarily exposed bedrock, the resulting channel flow line is flat in cross section. The average grade of the channel in the vicinity of the bridge is 0.3 percent.

The Obed River drains approximately 19.3 square miles above the bridge site. The Corps of Engineers has estimated the

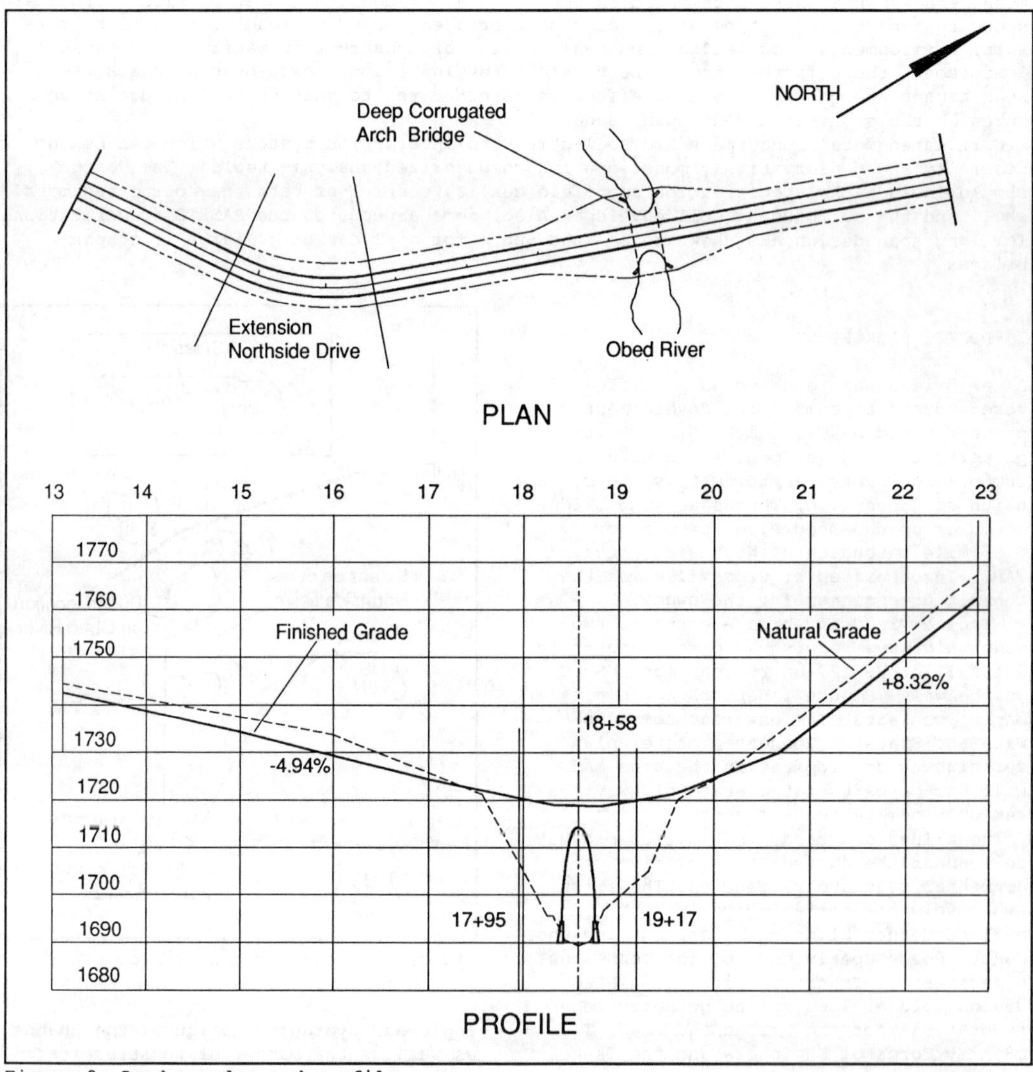

Figure 2. Roadway plan and profile.

maximum probable flood for this drainage area to be 2,500 cubic feet per second. Significant modification of the river channel and construction activities that would cause sedimentation downstream were restricted by the Corps of Engineers.

3 ALTERNATIVE BRIDGE TYPES

In preliminary studies for the bridge site the consultant considered standard, multiple barreled culvert alternatives for the application. Corrugated metal pipes, precast concrete boxes, and cast-in-place multiple cell concrete boxes were evaluated at various early stages of the design. However, these preliminary plans were rejected by the county planning department and the Corps of Engineers. While providing adequate hydraulic capacity, these low clearance alternatives resulted in excessively steep road grades. They also significantly impacted the bed of the natural channel. In addition, the Corps would not approve plans requiring diversion of the river during construction. Therefore, the only acceptable alternative was a bridge that would span the waterway.

Two bridge alternatives were then investigated that would meet the final design requirements described above. They were a prestressed concrete girder bridge and a corrugated metal long span culvert. After an economic evaluation, the long span product was chosen as the least cost alternative. Syro Steel Company of Girard, Ohio was chosen as the successful vendor for the long span bridge. Their 15-inch pitch by 5.5-inch deep corrugated product, known as Deep Cor, would be furnished in an 8 gage (0.168-inch) wall thickness with a step beveled end configuration.

4 ECONOMIC COMPARISONS OF BRIDGE TYPES

Cross sections of the two bridge alternatives have been depicted in Figures 3 and 4. The prestressed concrete girder bridge was to have an 87-foot clear span and be supported by piling at the poured-in-place concrete abutments. The precast vendor also specified riprap slope protection to prevent erosion and undercutting of the abutment slopes. Cost data for this alternate was furnished by the owner.

The corrugated plate alternate was to have a 40-foot, 9-inch span and a 19-foot, 9-inch rise. The multi-radius arch was to support 4.5-feet of cover and an HS20-44 live load. Footings, depicted in Figure 5, were designed by the owner's engineer with keyed embedment into the bedrock flowline of the riverbed. End treatment consisted

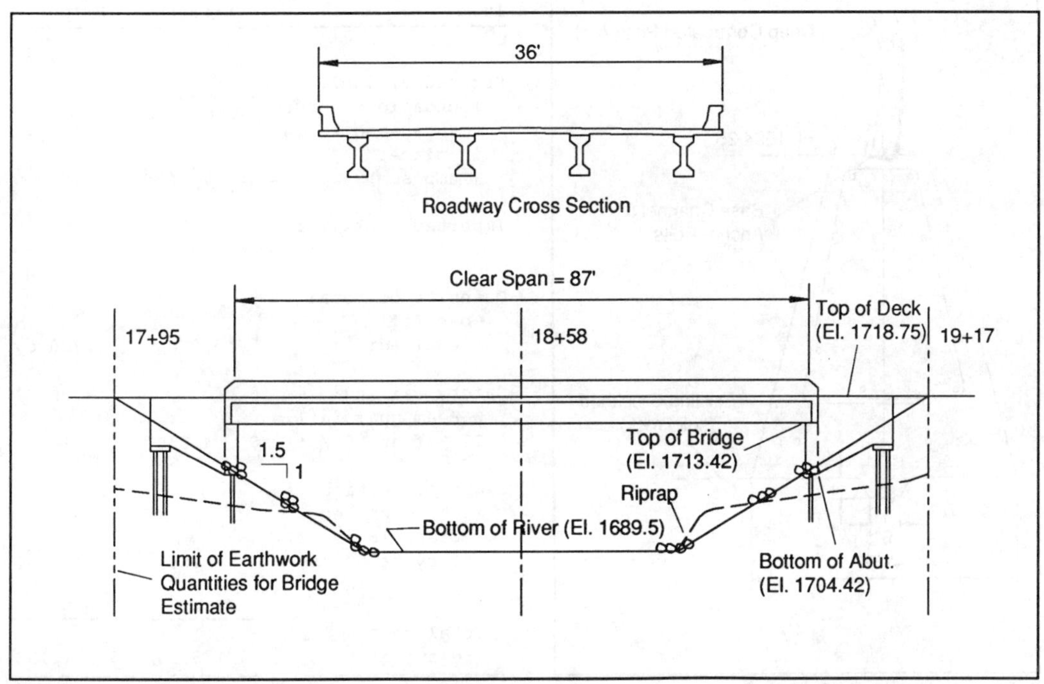

Figure 3. Concrete girder bridge alternative.

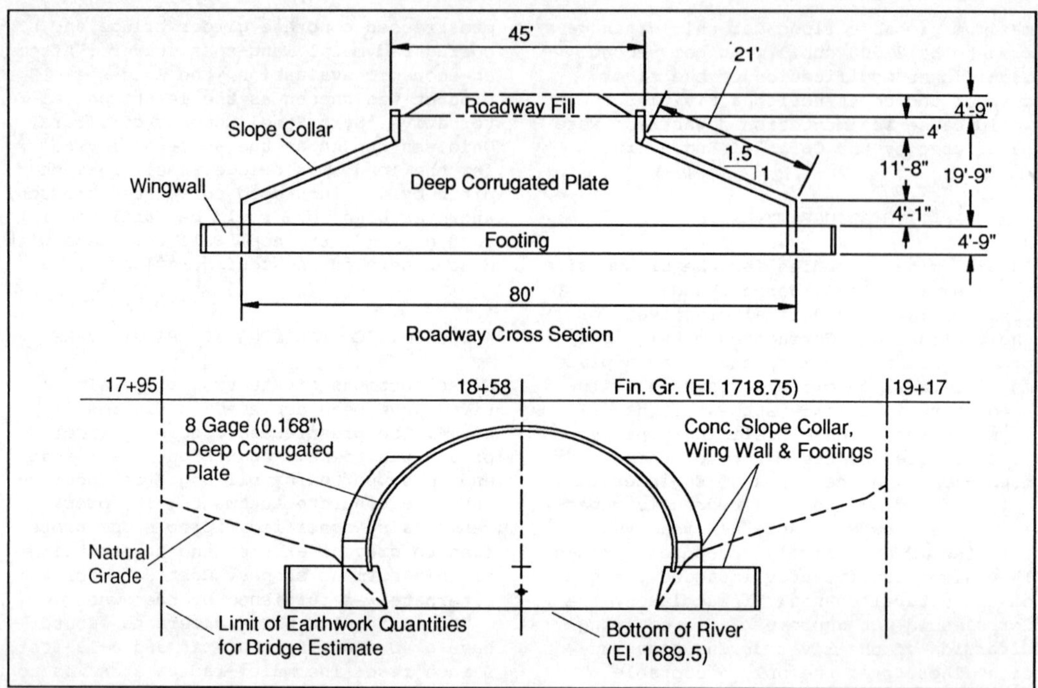

Figure 4. Deep corrugated long span alternative.

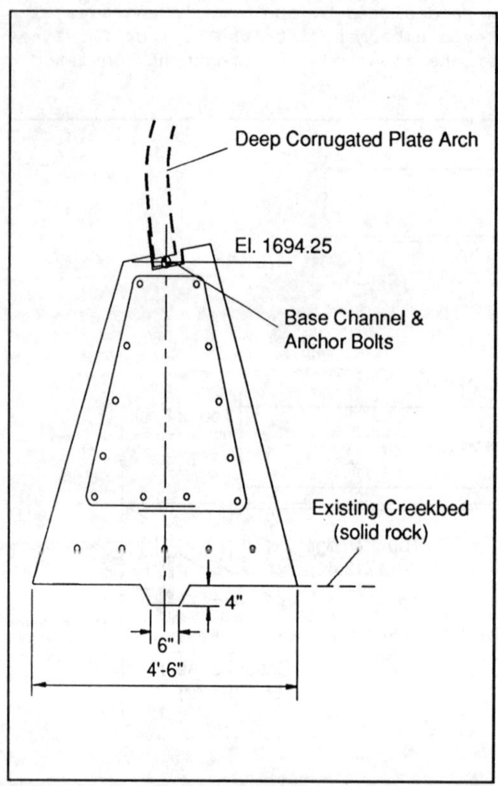

Figure 5. Typical footing cross section.

Table 1. Relative cost comparison of bridge alternatives.

Item	Long Span Bridge	Girder Bridge
Structure, incl. foundations, abutments, end walls, in-place: (Lump sum)	$ 77,550	$ 115,000
Imported backfill:	$ 54,000	$ 20,400
Quantity of total backfill required:	4,500 cy	1,700 cy
Safety railing: W-Beam guardrail, 174 lf at $25/lf	$ 4,350	incl
Slope protection: Machine placed riprap, 18-in tk, 350 sy at $45/sy	incl	$ 15,750
Total cost (relative)	$ 135,900	$ 151,150

54

of wing walls, slope collars, and header
walls. Therefore, no additional slope
protection was required for this alternate.

A cost comparison of the two alternatives has been given in Table 1 and was
based on 1988 dollars. Only relative cost
factors have been included. For instance,
the earthwork estimate was restricted to
quantities between Station 17+95 and Station 19+17 for both alternatives (see
Figures 3 and 4). Also, road paving, which
would be required for both alternatives,
was excluded from the estimate. The corrugated steel long span was shown to be the
lowest cost alternative by over $15,000.

Because of the steep terrain and minimum
slope requirements imposed by the county
planning department, the opportunity did
not exist to balance cut and fill over the
entire roadway project in accordance with
normal practice. As a result, a large
quantity of off-site borrow material was
required to complete the project. Normally, only about 1,800 cubic yards of
imported granular material would have been
required within the structural backfill
envelope for the long span alternative.
The remainder of the 4,500 total cubic
yards of fill required could have been on-
site material that could have been placed
and compacted at a much lower cost than
that of the borrow material.

The estimate shown in Table 1 assumed
the worst case for backfill costs by including 100 percent of the fill required
to be expensive imported material. This
negatively impacted the relative cost of
the long span alternative since long spans
generally require a larger percentage of
fill material than girder bridges. During
actual construction, however, a sizeable
quantity of stockpiled site material was
ultimately utilized within the limits of
the estimate but outside of the structural
backfill envelope, thereby lowering the
actual cost of construction below this
estimate. However, specific quantities
were not available. Hence, the worst case
has been reported.

Also, the estimate was prepared using
present cost analysis techniques. This
was justified since the maintenance-free
life expectancy of both alternatives
was projected to be far beyond the service
life requirements of most regulatory
agencies (50-years for bridges on the
Federal highway system).

In the case of the corrugated steel
long span alternative, the backfill material was tested and found to have the
properties shown in Table 2. Using recognized techniques, the life expectancy of
the steel structure was projected to be

Table 2. Backfill soil chemical properties.*

Natural Moisture:	10.0 %
Resistivity:	141,000 ohms/cc
pH:	5.30

* Test results per Geotek Engineering Co.,
Nashville, Tennessee.

195 years (AISI, 1983). This assumes a 25
percent metal loss, which is conservative
since the methodology was based on invert
corrosion of factory-made pipe. Hot-dip
galvanized structural plate has a minimum
of 50 percent more zinc coating than factory-made pipe. Also, the arch application
here does not have an invert, and the
design flow is totally contained within
the concrete footer abutments.

5 ARCH DESIGN METHODOLOGY

Corrugated structural plate long spans
mobilize the surrounding backfill material
into a soil-structure interaction system.
The constituents, a corrugated shell
and the soil envelope, depend upon one
another for support. Therefore, they must
be designed as a composite. But since the
resulting structure is highly indetermi-
nate and can only be characterized by
nonlinear engineering properties, these
systems defied rigorous structural analysis for many years. As a result, an empirical basis developed for their design.
Such a method is currently contained in
the AASHTO bridge design specification for
long spans.

However, a number of limitations exist
in the empirical method. Its application
is confined to the range of sizes, shapes,
and loadings originally included in the
development of the method. And, while most
follow-up studies have found the AASHTO
method to be conservative for its range of
application (small to moderate long span
sizes under typical highway loads), it
does not have the ability to assess the
effect of long term shape change due to
poor backfill. Also, it does not currently
provide a means of evaluating the required
size of the backfill envelope. However,
with the advent of the computer, a more
rigorous and general analysis tool for
soil-structure interaction became available.

Computerized analysis tools for soil-
structure interaction systems have been
developing for 20-years and currently

enjoy a substantial base of verification
(Duncan 1979, Katona etal. 1979, Chang
etal. 1980, McVay and Selig 1982, Leonards
etal. 1982, Selig and Musser 1985). These
tools have overcome most of the limitations described above for the empirical
method. Additional advantages include
the ability to evaluate arbitrary loading
and geometry conditions. Also, a greater
spectrum of structural response characteristics becomes available through their
use. This ultimately leads to more rigorous and reliable designs. For these reasons, a computerized design approach was
chosen for this project in lieu of the
empirical AASHTO methods. However, AASHTO
safety factors and general material specifications were fully complied with. The
detail design calculations for the arch
were furnished by the manufacturer.

The 40-foot, 9-inch span by 19-foot, 9-
inch rise, low profile arch shape was
analyzed using the CANDE finite element

Table 3. Hyperbolic parameters for soil
types used in the CANDE analysis.

Parameter	Backfill Soil	In Situ Soil
CANDE Soil ID:	CA105	SM100
Unit Weight (pcf):	145.00	120.00
Cohesion Intercept (psi):	0.00	0.00
Friction Angle (deg):	42.00	36.00
10-Fold Reduction in Friction Angle (deg):	9.00	8.00
Failure Ratio:	0.70	0.70
Modulus Number:	600.00	600.00
Modulus Exponent:	0.40	0.25
Bulk Modulus Number:	175.00	450.00
Bulk Modulus Exponent:	0.20	0.00

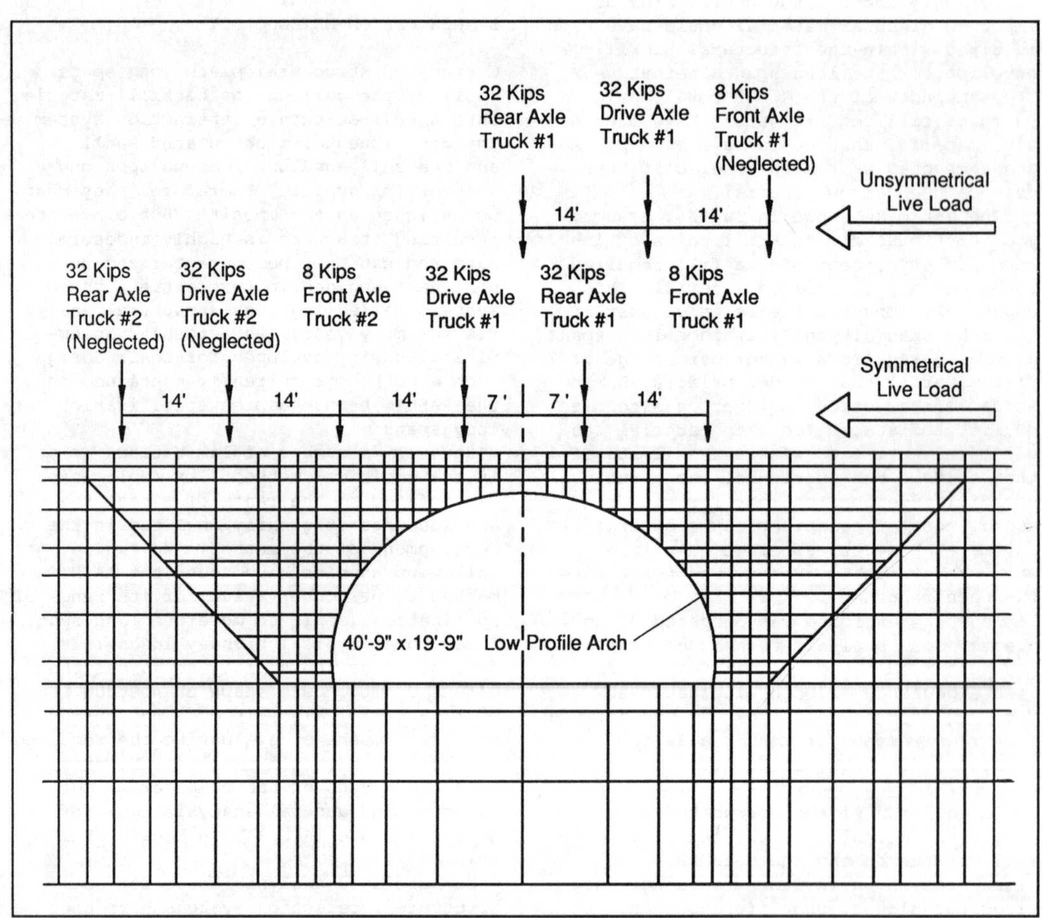

Figure 6. Finite element computer model of long span arch.

computer program. CANDE was developed
under sponsorship of the Federal Highway
Administration. It models soil-structure
interaction by means of incremental construction, interface slip between the
conduit wall and the soil, hyperbolic
stress-strain relationships for the soil,
and compaction induced pressures during
construction (Katona etal. 1976, Musser
1989). Versions of the program are available to run on mainframe, minis, and micro
computers.

Default soil parameters for the Duncan
hyperbolic model were readily available
in the CANDE program. (Katona 1981) Also,
previous experience had shown them to be
generally conservative in representing
properties of the specified soil types.
Therefore, with preliminary soils information, the Duncan CA105 and the Duncan
SM100 were chosen to represent the granular backfill and in situ materials respectively. Hyperbolic parameters for these
soils are given in Table 3.

Construction loads were simulated by
placing soil in five increments on the
side and three increments over the top of
the arch. A D-4 Cat dozer was modeled in
each load increment over the top. A one
pound per square inch compaction pressure
was also applied to the backfill material
at the sides of the arch using a "squeeze"
technique proposed by Katona etal. (1979).

Three loading combinations were investigated as follows:

1. Dead loads and construction live
loads induced during the backfill process.

2. One above plus symmetrical HS20-44
live load placed at design cover height.

3. One above plus unsymmetrical HS20-44
live load placed at design cover height.

Concentrated point loads of the HS20-44
vehicle configuration were represented by
average line loads (axle weight divided by
lane width). Symmetrical and nonsymmetri-
cal loading positions are shown in Figure
6, along with the finite element mesh employed. Note that vehicle loading positions falling outside the span of the
structure were conservatively neglected.

The conduit wall was modeled as an
elastic beam placed at the neutral axis of
the deep corrugated wall section. Geometry
and section properties of the 8 gage
(0.168-inch) corrugated product have been
given in Figure 7. Chemical, physical,
geometrical, and coating properties met
the requirements of ASTM A-761 (AASHTO M-
167).

The analysis demonstrated that the
greatest bending moments occurred during
construction just before backfill was
placed over the top of the structure. The
maximum moment at this stage of loading
occurred at the quarter span for this
configuration, which was deflecting inward. At this same time the crown was deflecting upward. This pattern of motion
was induced by compaction pressures being
exerted along the sides of the arch. The
safety factor against the formation of a
plastic hinge at this temporary construction interval was 1.82. This included the
interaction of both moment and thrust in a

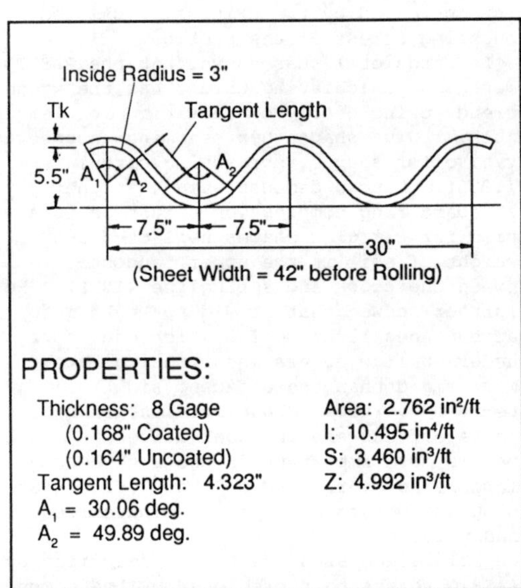

PROPERTIES:

Thickness: 8 Gage   Area: 2.762 in²/ft
(0.168" Coated)   I: 10.495 in⁴/ft
(0.164" Uncoated)   S: 3.460 in³/ft
Tangent Length: 4.323"   Z: 4.992 in³/ft
A_1 = 30.06 deg.
A_2 = 49.89 deg.

Figure 7. Deep corrugated plate section
properties.

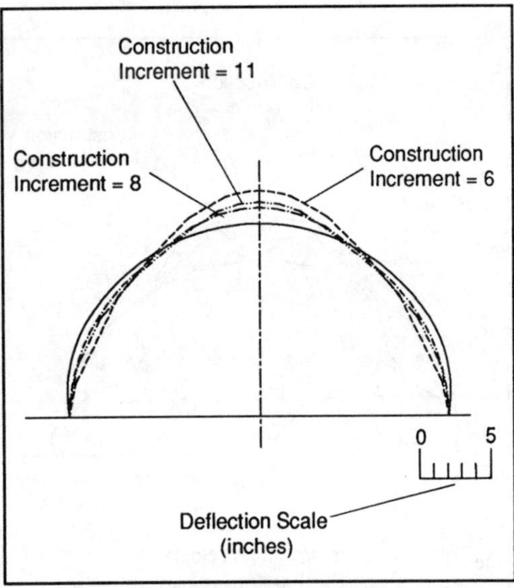

Figure 8. Net deflection diagram.

manner proposed by Leonards (1982).

Controlled upward peaking during backfill is both typical and desirable of flexible arch structures. It pretensions the top arc so that when loads are placed over the top, bending moments actually decrease. The analysis bore this out. Figures 8 and 9 depict the deflection and bending moment diagrams predicted by CANDE at various stages of loading. With backfill pressures only on the side, as in construction increment 6, the analysis predicted that a maximum crown peaking of 2.25-inches would occur.

With partial cover ramped over the top, as in construction increment 8, crown peaking was reduced to 1.30-inch and the minimum plastic hinge safety factor increased to 2.49. Finally, with backfill placed in the saddles up to finished grade and under the action of live load, as in construction increment 11, 1.56-inch of crown peaking still existed and the minimum safety factor against plastic hinge formation moderated to 2.06.

Also, at construction increment 11 the bending moments were nearly equal at the crown and quarter point for this configuration. The symmetrical loading case was only slightly more severe overall than the unsymmetrical case.

Choice of the compaction pressure magnitude used in the analysis was estimated, based on past experience. It was known that structure peaking during backfill was a function of structure size, geometry, stiffness, soil type, soil stiffness, thickness of compacted soil lift, compac-

tion equipment weight, type, and pattern of operation. Seed etal. proposed an analytically based modeling technique that could account for these factors (1983). However, in order to make use of his method, detailed assumptions regarding the magnitude of all the various input parameters listed above must be made ahead of time. At the time this project was evaluated, a generalized basis for these assumptions did not exist. Therefore the method was not useful for design.

By contrast to the computerized approach described above, the empirical AASHTO design method for long spans attempts to achieve a safe moment design indirectly by specifying geometric constraints on the shape and minimum wall stiffness via a table of minimum thicknesses for the 6-inch by 2-inch steel corrugated structural plate product. However, a quantitative measure of the bending stress safety factor is not available. The primary design criteria in the AASHTO method continues to be ring compression.

The calculation of ring compression for long spans in AASHTO is performed by multiplying the crown pressure by the top arc radius. The crown thrust is assumed to be constant around the periphery. Also, no criteria exists for consideration of more than one point of live load contribution, which is assumed to act at the crown. A safety factor of three is then applied to the ultimate capacity of bolted seams. A safety factor of two is applicable to wall sections where bolted seams do not occur. Here the ultimate capacity is controlled by the critical localized buckling stress of the section.

Leonards etal. has shown that the AASHTO method of calculating thrust has the wrong trend, being overly conservative for large multi-radius shapes but becoming unconservative for shapes approaching circular (1985). Duncan demonstrated that the standard ring compression thrust calculation for circular shapes neglected the weight of soil in the upper haunches between the crown and springline (1979). He further showed that it did not take into account negative arching which can occur under shallow covers when footings are more rigid than the adjacent side fill in terms of vertical consolidation. Both of these effects are unconservative and become more important as the arch becomes larger. However, both of these effects are programmed into a complete computerized analysis.

Table 4 compares the thrust specified by AASHTO versus that predicted by the computerized analysis performed for this proj-

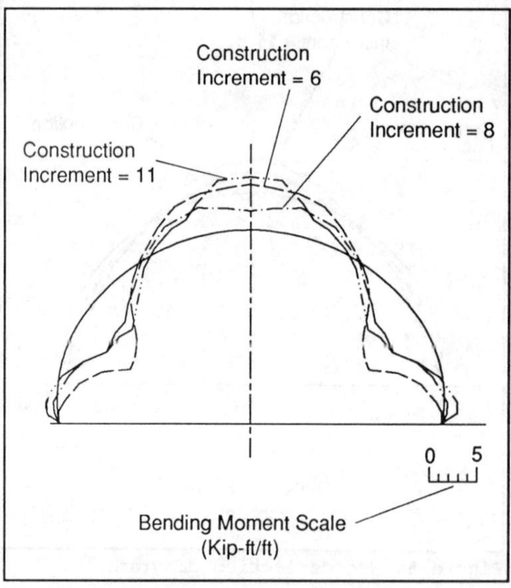

Figure 9. Bending moment diagram.

Figure 10. Erection of deep corrugated plate rings.

Figure 11. Backfill placement over top of arch.

ect. Note that the computerized analysis predicts a variable thrust around the periphery, with the maximum occurring near the springline for this configuration. Also, the AASHTO method under predicts thrust at all points around the periphery. This is thought to be a result of the factors described above, since the shape is very nearly a half-circle.

Another contributing factor for this difference between the two designs is likely the method of modeling live loads. Unlike the single point of application (at the crown) assumed for AASHTO culvert design, the computerized method evaluated regular HS20-44 load patterns in multiple placement scenarios, as shown in Figure 6. This generalized method is more typical of AASHTO requirements for conventional bridge types.

In terms of wall thrust, the 8 gage (0.168-inch) deep corrugated steel product provided a 3.03 safety factor against the worst case loading at a bolted seam, which

Table 4. Comparison of calculated thrusts.

Location	AASHTO Thrust[1]	CANDE Thrust[2]
Crown:	20,700 plf	21,300 plf
Lowest Bolted Seam (11'-3" above base):	20,700 plf	26,700 plf
Springline (2'-3" above base):	20,700 plf	30,200 plf

[1] Based on 22-foot top radius, 145 pcf soil density, and single 32 kip axle (typical HS20-44 representation).
[2] Based on 145 pcf soil density, multiple HS20-44 load patterns as shown in Figure 6, symmetric case.

occurred at 11-feet, 3-inches above the base of the arch. It also provided a 3.02 safety factor against the worst case wall compression stress, which occurred at springline, 2-feet, 3-inches above the base. Thus, the spirit of the AASHTO specifications was met or exceeded in all cases. Also, it should be noted that the above safety factors were based on the conservative assumption of 145 pounds per cubic feet soil unit weight. Actual site tests showed field densities fell in the range of 120-130 pounds per cubic feet.

6 CONSTRUCTION PHASE

Foundation piers were constructed in a keyway blasted into the bedrock foundation along the bank on each side of the river. Steel base channel was cast into the top of the foundation piers. The corrugated steel arch was then erected in rings on the ground and hoisted into place, as shown in Figure 10.

Erection of the arch was achieved with 174 man-hours using a four-man crew. Equipment included an 18 ton P & H crane, a snorkel man-lift, an air compressor and a pick-up truck with miscellaneous tools.

The import fill material was a silty sand intermixed with soft sandstone fragments which was extracted from a nearby borrow pit. Gradation consisted of 100 percent passing a 3/4-inch sieve, 79 percent passing a No. 4 sieve and 30 percent passing a No. 200 sieve. The liquid limit was 24.9 and the plasticity index was 6.3. The AASHTO classification was A-2-4 (AASHTO M-145, ASTM D-3282). Standard proctor density (AASHTO T-99, ASTM D-698) was 112 pounds per cubic foot at 13 percent optimum moisture. Material

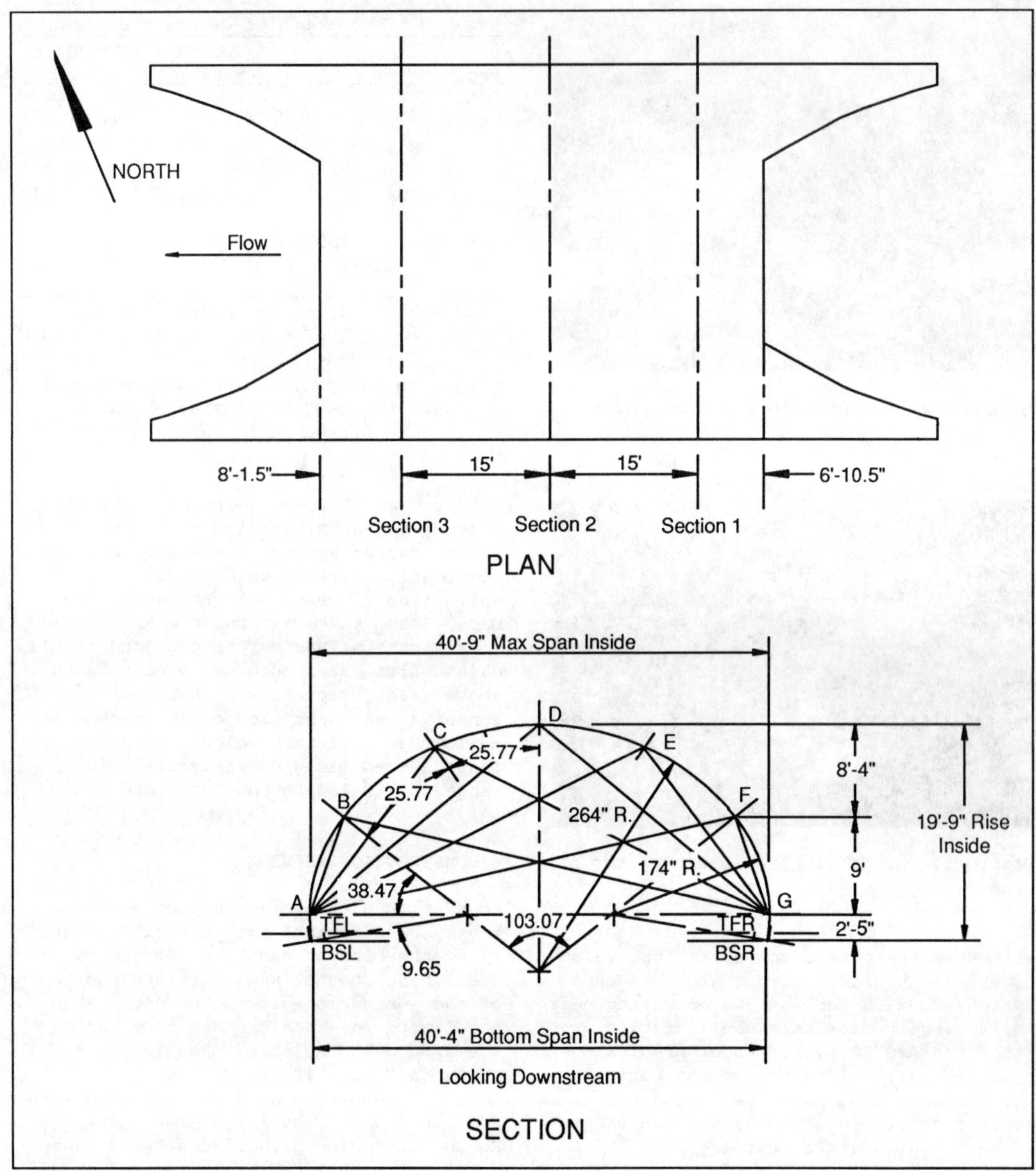

Figure 12. Shape monitoring stations.

was placed in 6-8 inch lifts and compacted by track rolling with a CAT 955 loader. Field density and moisture content was checked with a nuclear density meter (ASTM D-2922 and D-3017). Compaction effort was applied until 95 percent of the maximum proctor density was achieved. Laboratory and field soil testing was performed by Geotek Engineering Company of Nashville, Tennessee.

Backfill lifts were placed uniformly on both sides of the structure until the depth of backfill reached approximately 75 percent of the rise. At this point backfill was ramped over the top of the arch with a Cat D-3 bulldozer as shown in Figure 11. Also seen in this figure are the temporary supports required for the beveled ends of the arch. Temporary supports are required to maintain shape on beveled and skewed ends of long spans until permanent slope collars or header beams are in place.

In order to evaluate the effectiveness

Figure 13. Live load test.

of the analysis in predicting field behavior, extensive deformation monitoring
was performed during construction on the
arch. Three cross-sections were chosen and
nine control points were established
around the periphery at each cross-section
as shown in Figure 12. Measurements were
taken at various stages during backfill
placement.

The maximum peaking observed was 3.12-
inches. It occurred just before backfill
was placed over the top and amounted to
about 40 percent more than predicted. This
means that the pretensioning achieved, and
hence the safety factor against future
crown overloads, were greater than predicted using the compaction modeling
technique described above.

Shape monitoring was continued until
backfill reached approximate finished
grade elevation, 55-inches over the crown.
At this point and before pavement construction a live load test was performed
using a heavy duty dump truck with rear
dual axles centered over the crown at
measuring station number 2 (see Figures 12
and 13). The weight of the loaded truck
was measured on the scales of the local
farmer's co-op. Gross vehicle weight was
found to be 82,540-pounds while the rear
tandem axles weighed 63,600-pounds, almost
twice the design HS20-44 axle load specified.

Structural shape measurements before and
after placement of the live load detected a crown deflection of 0.18-inch.
The computerized analysis predicted a
live load crown deflection of 0.23-inch
under the standard HS20-44 load (see
Figure 6). This conservatism in the analysis was thought to be due to soil model
chosen which was known to be softer than
reality.

Total construction time from completion

of foundations to completion of subgrade
elevation required 15 working days. This
included erection of the arch and backfill
to a height which permitted crossing of
highway rated construction equipment.

7 CONCLUSIONS

1. Many environmental and site factors
enter into the economic comparison of
bridge alternatives. These factors include
road profile, site terrain, environmental
regulation, and availability of construction materials. To assure least cost,
these factors should be carefully evaluated on a case-by-case basis.

2. Corrugated metal long spans are shown
to be cost effective alternatives to conventional bridges in the small to medium
span range. They should be evaluated as
viable alternatives for new bridge construction or bridge replacements on the
nation's highway system.

3. Computerized analysis tools can be
used to successfully and conservatively
predict the performance of soil-structure
interaction systems. It is important to
account for all factors affecting structural performance including the size and
characteristics of the soil envelope,
compaction pressures, and soil-conduit
interface behavior.

4. The AASHTO specification for long
span design is shown to be less rigorous,
and sometimes unconservative, for large
long span structures. Large long span
bridges should be designed with the same
rigor and care as conventional bridges.

5. Careful attention must be paid the
design and construction of the soil envelope. In addition to structural shape
monitoring during construction, quality
control of backfill material and placement
is essential to the successful performance
of a long span installation.

ACKNOWLEDGEMENTS

Thanks are due the owner, William F.
Graham, for his cooperation in the development of this study.

REFERENCES

AISI. 1983. Handbook of Steel Drainage &
    Highway Construction Products. American
    Iron and Steel Institute. Washington,
    D.C.
Chang, A.M., Espinoza, J.M., and Selig,
    E.T. 1980. Computer Analysis of Newtown

Creek Culvert. Journal of the Geotechnical Engineering Division, ASCE GT5:
531-556.

Duncan, J.M. 1979. Behavior and Design of
Long-Span Metal Culverts. Journal of the
Geotechnical Engineering Division, ASCE
GT3: 399-418.

Katona, M.G., Smith, J.M., Odello, R.S.,
and Allgood, J.R. 1976. CANDE - A Modern
Approach for the Structural Design and
Analysis of Buried Culverts. Report No.
FHWA-RD-77-5. Federal Highway Admini-
stration. Washington, D.C.

Katona, M.G., Meinhert, D.F., Orillac, R.,
and Lee, C.H. 1979. Structural Evaluation of New Concepts for Long-Span Culverts and Culvert Installations. Report
No. FHWA-RD-79-115. Federal Highway
Administration, Washington, D.C.

Katona, M.G., Vittes, P.D., Lee, C.G., and
Ho, H.T. 1981. CANDE-1980: Box Culverts
and Soil Models. Report No. FHWA-RD-80-
172. Federal Highway Administration.
Washington, D.C.

Leonards, G.A., Wu, T.H., and Juang, C.H.
1982. Predicting Performance of Buried
Conduits. Report No. FHWA/IN/JHRP-81/3.
Indiana State Highway Commission.
Indianapolis.

Leonards, G.A., Juang, C.H., Wu, T.H., and
Stetkar, R.E. 1985. "Predicting Performance of Buried Metal Conduits. Transportation Research Board Record 1008:
42-52.

McVay, M.C. and Selig, E.T. 1982. Performance and Analysis of a Long-Span Culvert. Transportation Research Record
878: 23-29.

Musser, S.C. 1989. CANDE-89 User Manual.
Report No. FHWA-RD-89-169. Federal
Highway Administration. Washington,
D.C.

Seed, R.B. and Duncan, J.M. 1983. Soil-
Structure Interaction Effects of Com-
paction-Induced Stresses and Deflections. Report No. UCB/GT/83-06. University of California, College of Engineering. Berkeley.

Selig, E.T. and Musser, S.C. 1985. Performance Evaluation of a Rib Reinforced
Culvert. Transportation Research Record
1008: 117-122.

Structural Performance of Flexible Pipes, Sargand, Mitchell & Hurd (eds) © 1990 Balkema, Rotterdam. ISBN 90 6191 165 6

Field performance of steel corrugated, box-culvert with corrugated plate reinforcement

S.Oh, G.A.Hazen & S.M.Sargand
Center for Geotechnical and Groundwater Research, Ohio University, Athens, Ohio, USA

J.O.Hurd
Ohio Department of Transportation, Columbus, Ohio, USA

ABSTRACT: The structural behavior of a circumferentially stiffened, low profile, corrugated steel box culvert under shallow soil cover is investigated. Deflections and strains are monitored at the central section during construction and live loading. Measurements are compared to results from the Culvert ANalysis and DEsign (CANDE) and Soil Structure Interaction Two (SS12) finite element programs. Analysis for the effect of the live loads is determined with ILLI-SLAB. The finite element analyses of the early stages of backfill placement underestimate crown deflection and moment. Poor correlations of measured to analytical moments, and of measured to analytical deflections are observed for live load applications. The finite element analysis predicts higher values than those measured.

1 INTRODUCTION

Because low profile box culverts are designed for use under shallow soil cover, stiffening ribs are bolted to the crown and sides. This prevents the occurrence of excessive stress and deformation at the haunch and crown. The pressure of the surrounding soil contributes to crown deflection and attenuates lateral deflections.

Rib reinforced, long-span, and box culverts have been the subject of several studies in recent years. Duncan (1982, 1983) performed finite element analysis (FE) on rib reinforced, aluminum box culverts to develop design equations. He conducted full scale field tests to validate the Culvert ANalysis and DEsign (CANDE) computer program. His measured live-load deflections were only 25% as large as the CANDE results, and measured moments at the haunch were also significantly smaller than CANDE's predictions. Beal (1981, 1986) tested a similar aluminum box culvert in the laboratory. He reported that strains were inconsistent and the bolt line and thrusts could be neglected. The Lane Metal Products

Company (1981) performed a load test on a corrugated, reinforced, low profile, box culvert. The deformations and stresses under load were satisfactory, with no yielding, bolt failure, or plate slippage detected.

2 DESCRIPTION AND INSTRUMENTATION OF THE CULVERT

For this investigation, a reinforced, low profile, box culvert was constructed on 3 foot by 3 foot footers. Dimensions of the culvert are: length 40 feet, span 15 feet 9 inches, rise 5 feet, radius of haunch 48 inches, and crown 10 feet 5 inches across. The culvert was constructed of 6 inch by 2 inch corrugated steel structural plate reinforced with circumferential corrugated steel stiffeners. Figure 1 shows a cross-section of the culvert. A total of 40 inches of cover over the crown was applied in the following layers: 20 inches compacted in 6 inch lifts, 8 inches of crushed limestone, and 12 inches of asphalt pavement.

Load was supplied by a dump truck

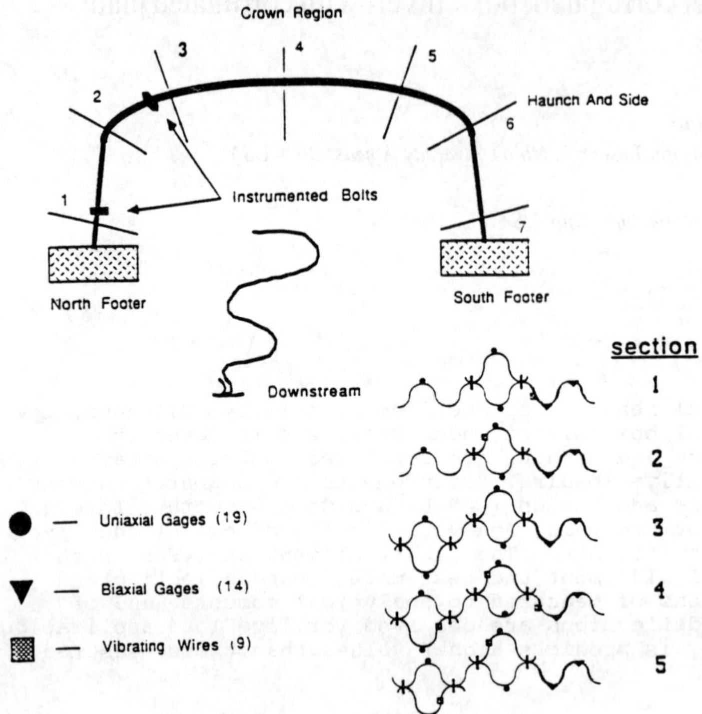

FIG. 1   CULVERT INSTRUMENTATION AND LOCATION OF GAGES

loaded with crushed limestone at 50, 100 and 130 percent of an HS-20 load. Stress was computed to be 16, 32 and 42 kips, respectively. Each load was placed over five positions on the culvert - four centered over long axis of the culvert with the front of the truck over the north end of the culvert, and the remaining one placed near the edge of the culvert with the front of the truck over the center of the culvert.

Eleven reference points were set up around the culvert periphery prior to backfilling. Horizontal and vertical deflections at each reference point were measured with a tape extensometer. Readings were taken at each stage of construction and when the culvert was subjected to live loads. Electric strain gauges were attached at the seven locations indicated in Figure 1, along a cross-section halfway down the length of the culvert. Soil displacement was measured with four rod extensometers: two located 10 inches above the foundations to measure spring line deflection and two others placed above the crown to measure arching. Two bolt areas were instrumented with a total of 27 strain-gauge rosettes. The configuration of sensors about a bolt is illustrated in Figure 2. Nine vibrating wire strain gauges were placed on the plate and ribs to verify the linear elastic behavior of the composite culvert system.

3   PREDICTION OF CULVERT RESPONSE BY FE METHODS

The finite element programs CANDE and Soil Structure Interaction Two (SSI2) were used in this investigation to predict culvert response to the construction sequence and live loads. The CANDE computer program was written in 1976. A detailed description of the program is in Katona (1976).

The SSI2 program, sometimes

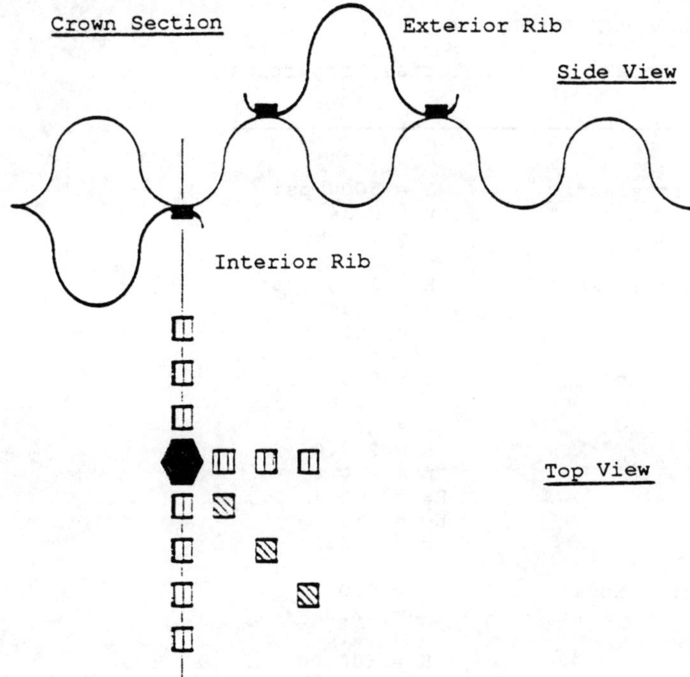

FIG. 2   ROSETTE ORIENTATION AT HAUNCH

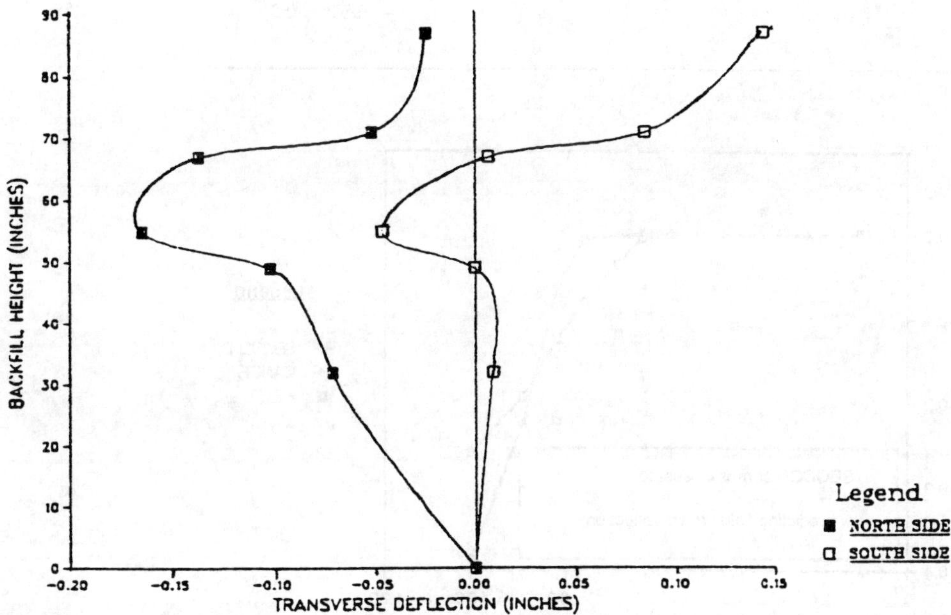

FIG. 3   HORIZONTAL DEFLECTIONS OF NORTH AND SOUTH SIDES

TABLE 1   SOIL MODELS AND MATERIAL PROPERTIES

Material	Soil Model	Material Properties
In -situ soil	Linear elastic	$E = 5000$ psi $v = 0.34$ $= 135$ pcf
Concrete footer	Linear elastic	$E = 3 \times 10^6$ psi $v = 0.34$ $= 150$ pcf
#603 backfill	Duncan Model	$c = 0.0$ $= 33.0$ $= 3.0$ $K = 200.0$ $n = 0.4$ $R_f = 0.7$ $K_b = 50$ $m = 0.2$
Limestone subgrade	Duncan Model	$c = 0.0$ $= 42.0$ $= 9.0$ $K = 600.00$ $n = 0.4$ $R_f = 0.7$ $K_b = 175$ $m = 0.2$
Asphalt	Linear elastic	$E = 3.5 \times 10^5$ $v = 0.34$ $= 150$ pcf

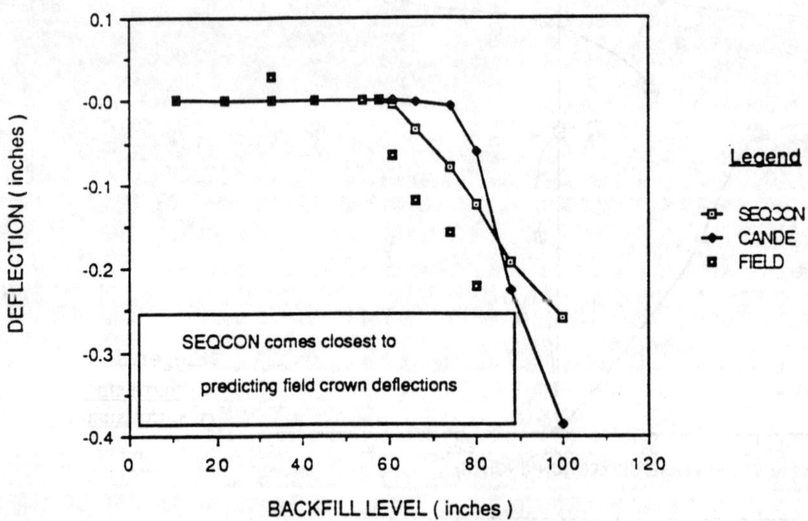

FIG. 4   FIELD AND FE CROWN DEFLECTIONS DURING BACKFILL

66

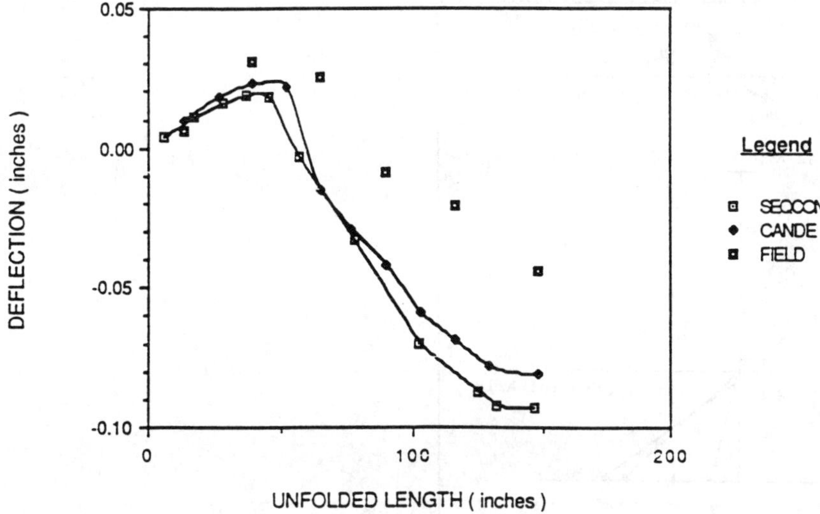

FIG. 5   FIELD AND FE DEFLECTIONS DUE TO 1.3 HS LIVE CENTER LOAD

referred to as SEQCON, uses eight-node isoparametric elements that allow non-constant stresses in each element in the mesh, which can be drawn to simulate a number of construction sequences. Nonlinear models for materials can be incorporated into SSI2. In this investigation, the Duncan or hyperbolic model was used to represent the granular backfill and subbase layers. Values of material properties as measured in the lab are presented in Table 1.

The ILLI-SLAB program was developed at the University of Illinois in the 1970's to model rigid pavements. In this investigation, it was used to back-calculate the response of the composite system to determine live load as used in CANDE.

RESULTS

The culvert crown exhibited upward deflection as the backfill was placed along the culvert sides. The maximum upward deflection was 0.029 inches, recorded after 29 inches of backfill had been added. The maximum inward deflection was 0.170 inches on the North side and occurred after 49 inches of backfill had been added. This is shown in Figure 3. The maximum downward deflection at the crown

was 0.24 inches, as shown in Figure 4. The deflections from the live load were much less; the maximum live load deflection was 0.149 inches and occurred when the rear truck wheels of the dump truck loaded to 130 percent HS-20 load was parked directly over the culvert center cross section. Figure 5 contains the results of this loading plus two simulations that will be discussed later.

Composite action of the sections was examined by plotting strains versus distance across the resisting section, as in Figure 6. Since the results are linear, the section response was analyzed as composite.

During the early stages of backfilling, negative moments developed in the crown and positive moments developed in the haunch and side sections. When backfill was placed over the crown, the signs of the moments reversed in the crown and haunch sections. The maximum negative moment after full covering of the culvert was 1.79 ft-kip/ft on the North haunch, while the maximum positive moment of 2.81 ft-kip/ft occurred at the crown. These data are shown in Figures 7 and 8. In contrast to Beal (1981, 1986), thrusts were measured to be significant. Maximum thrusts occurred at the North foundation and at the North crown at 8.14

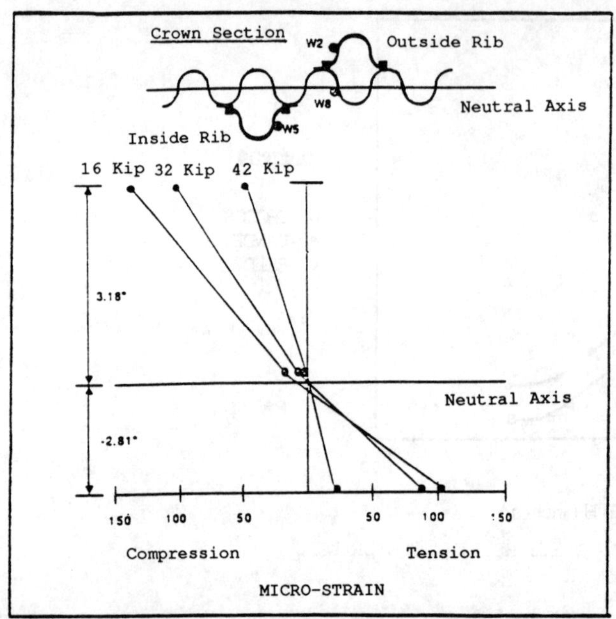

FIG. 6   VERIFICATION OF RIB-PLATE COMPOSITE RESPONSE

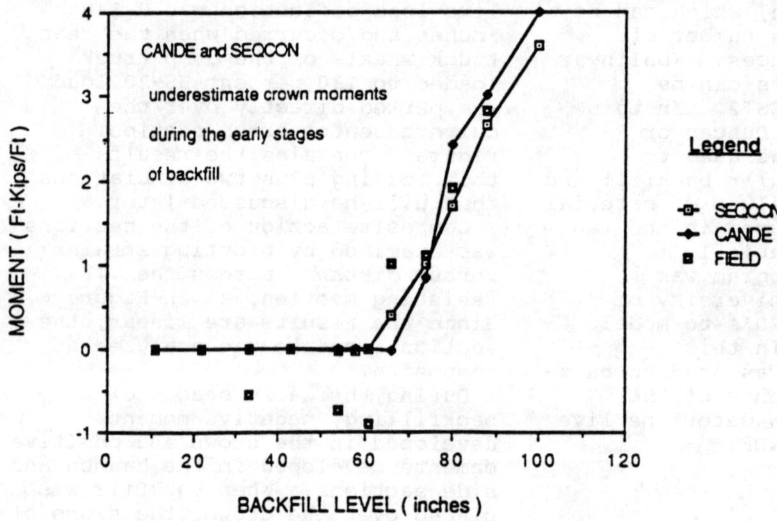

FIG. 7   FIELD AND FE CROWN MOMENTS DURING BACKFILL

kip/ft and 12.1 kip/ft, respectively. The occurrence of maximum moments and thrusts on the North side of the culvert is most likely due to an asymmetric backfill sequence.

The greatest moments during the live load measurements occurred when the truck wheels were parked directly above the culvert center. Positive moments resulted at the crown and foundation and negative moments developed at both haunches. The maximum moment was at the

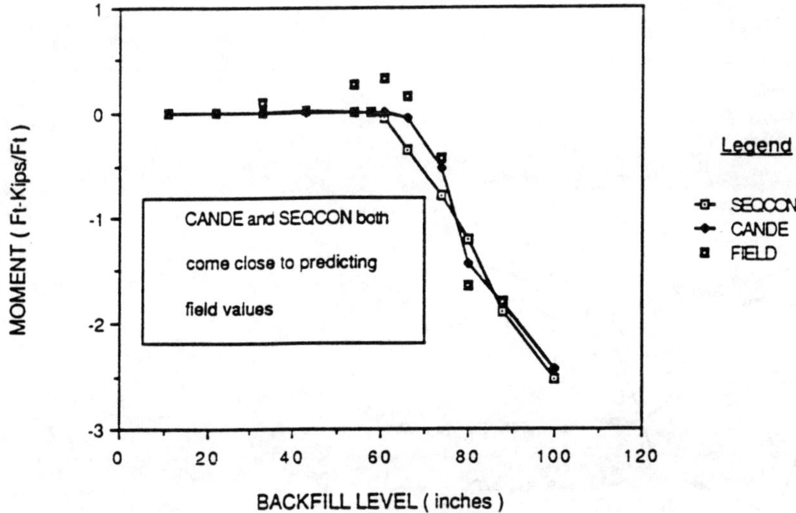

FIG. 8   FIELD AND FE MOMENTS HAUNCH DURING BACKFILL

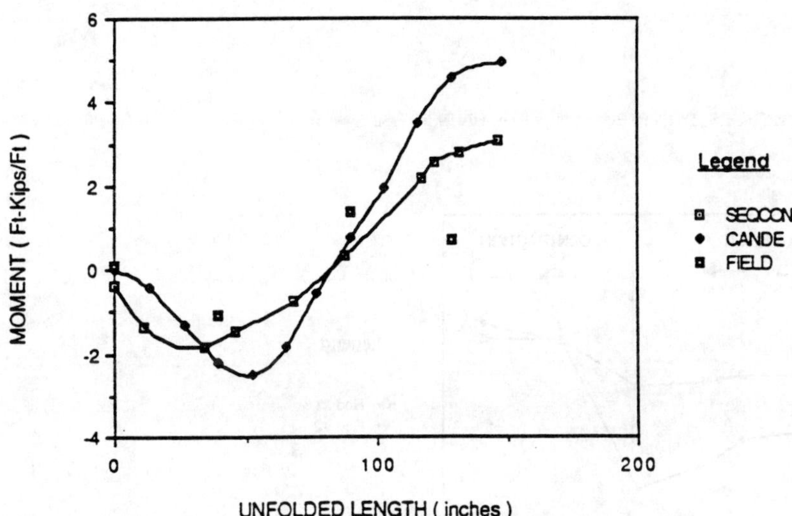

FIG. 9   FIELD AND FE MOMENTS DUE TO 1.3 HS LIVE CENTER LOAD

crown: 1.70 ft-kip/ft. Moments for all cross sections under the maximum live load (130 percent of HS-20 load) are shown in Figure 9. When the loaded wheels were parked directly over that side, the maximum thrust of 17.5 kip/ft was recorded.

Stress variations due to the bolts located at the haunch and footer were measured with patterns of electric strain gauge rosettes. Thrusts were large at the foundation, whereas the moments were large at the haunch. At the footer, stresses above the bolts during live loads were compressive. The large variation in stresses near the footer strongly indicated the presence of secondary effects. Figure 10 shows the results of measuring tensile stresses near the

69

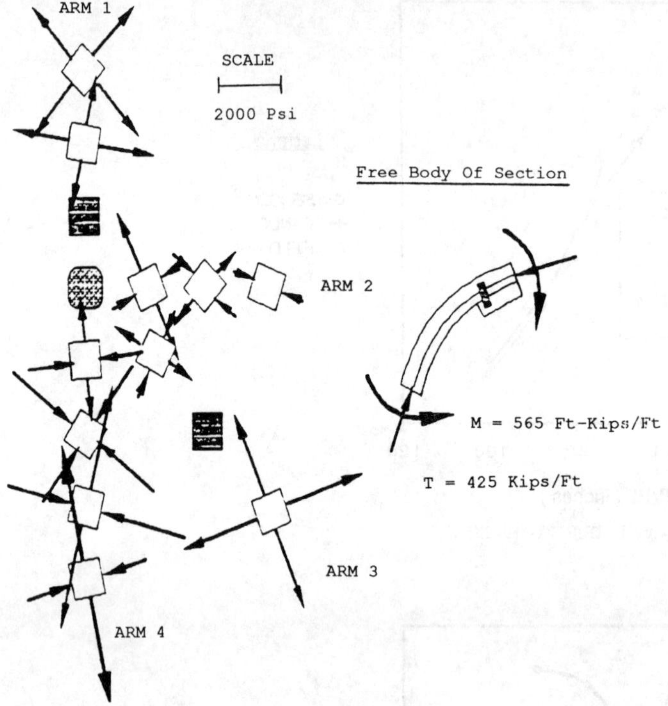

FIG. 10   PRINCIPAL STRESSES DISTRIBUTION AT A HAUNCH BODY
DUE TO 1.3 HS LOADING

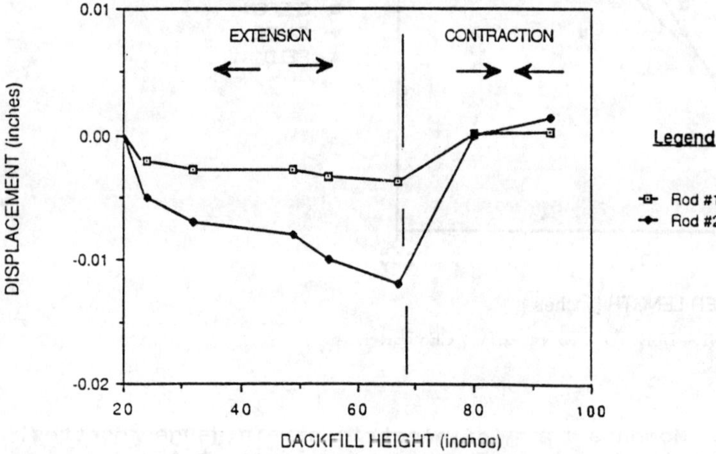

FIG. 11   DISPLACEMENTS OF SIDE RODS DURING BACKFILL

bolts at the haunch. The
asymmetric cross section was
expected to produce a rotation of
the bending stress axis, but this
could not be verified. Shearing
stress is shown by the orientation
of principal stress axes.

The soil extensometers located
near the foundation extended as
backfill was deposited at the sides
(Figure 11). As backfill was
placed over the crown, the culvert

shifted with a contraction in the
North rod and an extension of the
South rod. Both top rods extended,
indicating a negative arching.
Under application of live loads,
the side rods contracted and the
top rods extended.

CANDE and SSI2 were used for FE
of both the backfilling procedure
and the application of live load.
For analysis of the burying of the
culvert, both FE programs tended to
underestimate the magnitude of
deflection by a factor of about 2,
as shown in Figure 4. SSI2 gave
the more realistic results. On the
other hand, Figure 5 shows that the
FE programs tended to overestimate
the actual deflection under live
load by an even larger factor.
SSI2 and CANDE produced very
similar, mismatched estimates of
the deflection. However, CANDE and
SSI2 both gave reasonable results
for the behavior of crown moments
at later stages of backfill, but
underestimate the magnitude of the
negative crown moments observed in
early stages of backfilling. This
is shown in Figure 7.

The process of adding backfill to
the culvert was simulated in CANDE
by filling elements with soil in 11
layers in the FE, a technique that
also avoided numerical
instabilities involving nonlinear
constitutive laws. Unfortunately,
the effects of the compaction of
each layer could not be simulated.
The asymmetry in backfilling was
modeled by using two meshes, one
for each half of the culvert.
There was generally no significant
difference between the two methods
in the results for the deflections.
The same is true to a lesser extent
for the thrusts and moments, but
the lack of agreement between CANDE
and measured values persists.

CONCLUSIONS

The conclusions reached in this
field investigation were in
agreement with past investigations.
*   Moments and thrusts resulting
     from an application of live
     load were smaller than those
     resulting from backfilling.
     Thus, backfill was of good
     quality and compacted
     sufficiently.

*   Uneven backfill placement
     resulted in unsymmetric moment
     and thrust distributions
     around culvert periphery.
*   The finite element program for
     the design of metal culverts
     should be modified. First in
     culvert design CANDE predicts
     live load to be dominant when
     a box culvert is placed under
     2-5 ft. of cover. This was
     not true when compared to
     field values. Second, the
     design moment from CANDE is
     larger than that experienced
     by the structure.
*   This type of culvert, which
     uses a bolted corrugated rib
     as reinforcement for bending,
     can be considered composite
     for analysis purposes.
*   All stresses measured in the
     structure or near the bolts
     were well within design
     limits.

REFERENCES

Duncan, J.M. March 1982. Behavior
  and design of long-span metal
  culverts. Report No. UCB/GT/82-
  09, Dept. of Civil Engineering,
  University of California,
  Berkeley.
Duncan, J.M. and Drawsky, R.H. May
  1983. Design procedures or
  flexible metal culvert
  structures. Report No.
  UCB/GT/83-02, Dept. of Civil
  Engineering, University of
  California, Berkeley.
Beal, D.B. June 1981. Behavior of
  an aluminum structural plate
  culvert. Research Report 90,
  Engineering Research and Trans.
Gorman, C.D. April 1981. A
  comparison of field results with
  the theoretical analysis of Lane
  Metal Product's low-profile box
  culvert. Report No. 76-67-1,
  Technical Service Division,
  Engineering Dept., Bethlehem
  Steel Corporation.
Katona, M.G., et al., October 1976.
  CANDE: A modern approach for the
  structural design of buried pipe
  culverts. Report FHWA-RD-77-5.

Structural Performance of Flexible Pipes, Sargand, Mitchell & Hurd (eds) © 1990 Balkema, Rotterdam. ISBN 90 6191 165 6

Spiral rib metal pipe structural performance limits

James C. Schluter
Contech Construction Products, Inc., Middletown, Ohio, USA

ABSTRACT: Spiral Rib metal pipes were developed in the mid 1980's. They represent a unique wall configuration for metal pipes developed to provide flow characteristics equal to those for piping systems normally considered "smooth wall". To facilitate design on a conventional ring compression basis, the International Spiral Rib Pipe Association (ISRPA) has developed properties and limits for design. These properties, limits and their basis are provided here.

1. INTRODUCTION

Manufactured from either steel or aluminum, Spiral Rib metal pipes use a unique ribbed wall profile to provide stiffness while maintaining smooth flow characteristics. The pipe wall is spirally formed and wrapped using rectangularly formed ribs between flat wall areas as shown in Figure 1. Currently, two configurations (3/4 x 3/4 x 7-1/2 and 3/4 x 1 x 11-1/2) are offered for buried piping systems (Figure 2). These configurations provide pipes from 18" to 108" diameters. However, the technology lends itself to deeper rib patterns and larger diameter pipes for the future.

The combination of relatively wide flat areas interrupted by narrow ribs provide essentially smooth flow (Figure 3) and measured Mannings roughness factors (n-values) of 0.011 to 0.013. However, from a structural view, the width to thickness relationship of these flat elements does not allow the pipes to develop the full yield strength of the metal. To facilitate design, the uneffective portions of the pipe wall, which essentially is a "noncompact" section, needed to be accounted for. As with any new pipe wall configuration, flexibility factor (FF) limits also needed to be established to ensure installation stiffness.

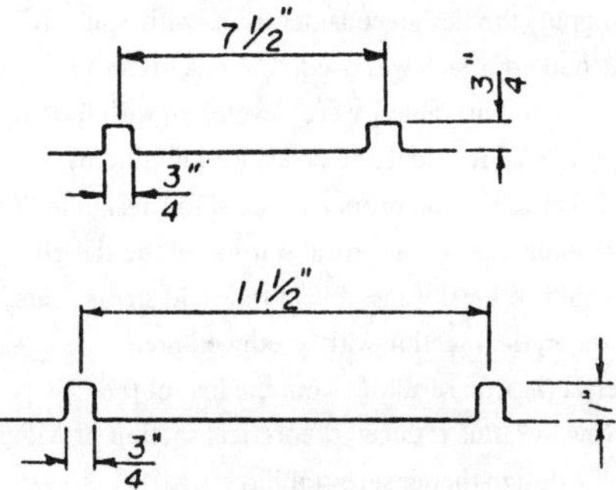

Figure 2. Spiral Rib Pipe Wall Configurations

Figure 1. Spirally Wrapped Spiral Ribbed Pipe

Figure 3. Smooth Inside Barrel of Spiral Rib Pipe

73

2. SECTION PROPERTIES

It's long been recognized that relatively wide, thin elements as represented by the flat portion of the Spiral Rib configuration do not function in compression as the geometrical section properties would suggest. Rather, if the section is "noncompact", the central portion of the flat element most remote from the stiffer rib, will buckle locally (a localized oil-canning effect) at a stress level that is below the yield stress. While the wall section as a whole has not buckled, the localized buckle renders the central portion of the flat element no longer available to carry compressive stress. Sections of this type perform well structurally as long as the critical compressive stress level is not reached. This stress level is dependent on the width-to-thickness relationship of the flat element (W/t) and the modulus (E) of the material.

Two design approaches can be used to ensure structural performance:

1. Limit the compressive design stress in the pipe wall so that the flat element is not stressed beyond its critical, buckling stress.

2. Develop effective section properties, by reducing them below the full geometric property levels such that the flat element will not buckle when full yield stress applied to the effective section.

To simplify the design considerations with Spiral Rib pipe, the second approach was used. The net, effective properties provided in Table 1 were developed with the full yield stresses (33 ksi for steel and 24 ksi for aluminum).

Net effective section properties used for design in this manner essentially remove a central portion of the flat element that would be locally unstable at the yield stress. This provides a theoretical section with a reduced area (A), moment of inertia (I), etc. resulting from the loss of the material. However, this reduced theoretical section provides properties for design that ensure stability.

When standard factors of safety for conventional metal pipes are applied, the designer is ensured that the full factor of safety is available to carry service loads if required.

Section properties were developed analytically using the light gage, cold formed metal design codes. For steel, the effective portion of the flat (b) can be determined by:

$$b/t = \frac{326}{\sqrt{f}}\left[1 - \frac{71.3}{(w/t)\sqrt{f}}\right] \dots\dots\dots\dots (1)$$

Where:

 b = Effective width of the flat element (in.)
 t = Thickness of the flat element (in.)
 f = Yield stress (33 ksi)
 w = Full width of the flat element (in.)

For aluminum Spiral Ribbed pipes a similar equation was developed using the modulus of aluminum (10,000 ksi).

As shown in Table 2, the effective properties are reduced more significantly for the lighter gages and for aluminum. This is as expected since the buckling stability of the flat element is dependent on its moment of inertia ($t^3/12$ for a flat element) and its modulus (E).

Several short column buckling tests were conducted as an additional check on the analytical procedure. Test results, accompanied with a comparison to analytical results are summarized in Figure 4 for the 3/4 x 3/4 x 7-1/2 aluminum configuration.

Once the effective area is determined, the remaining section properties are developed conventionally assuming the uneffective portion of the flat is not there. It must be remembered however that while the flat is in tension or under limited amounts of compressive stress it is functional and the full geometric properties are available.

Using the net effective section properties provides a conservative design approach. Both the analytical and short column tests use (assume) a straight section, unsupported by backfill. The curvature of the pipe wall, further supported by backfill, enhances the buckling stability of the central portion of the flat element. These effects have conservatively been ignored.

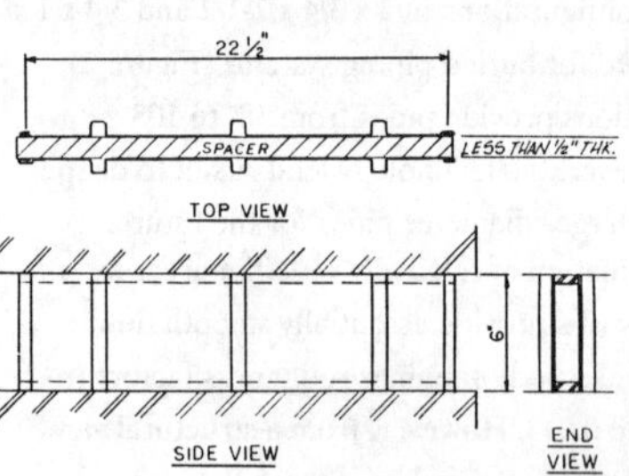

TEST CONFIGURATIONS

Effective Area (In^2/Ft.)		
Gage	Test Value*	Theoretical
.060	.416	.417
.075	.572	.572
.105	1.021	.922
.135	1.314	1.302

*Minimum value of 3 tests
Note: Tests performed by Dr. Ray Seed
University of California, Berkeley

Figure 4. Short Column Buckling Test Results
3/4 x 3/4 x 7-1/2 Aluminum
Test Results vs. Analytical Results

Table 1 - Net Effective Section Properties for Design

Steel Spiral Rib Pipes						
	3/4 x 3/4 x 7-1/2 Configuration			3/4 x 1 x 11-1/2 Configuration		
Thick-ness (in.)	A_s (in²/ft)	r (in)	I x 10⁻³ (in⁴/in)	A_s (in²/ft)	r (in)	I x 10⁻³ (in⁴/in)
.064	.511	.290	3.590	.374	.383	4.580
.079	.715	.282	4.740	.524	.373	6.080
.109	1.192	.268	7.150	.883	.355	9.260

Aluminum Spiral Rib Pipes						
	3/4 x 3/4 x 7-1/2 Configuration			3/4 x 1 x 11-1/2 Configuration		
Thick-ness (in.)	A_s (in²/ft)	r (in)	I x 10⁻³ (in⁴/in)	A_s (in²/ft)	r (in)	I x 10⁻³ (in⁴/in)
.060	.417	.303	3.200	.312	.396	4.080
.075	.572	.299	4.260	.427	.391	5.450
.105	.922	.290	6.460	.697	.380	8.390
.135	1.302	.284	8.740	1.009	.369	11.480

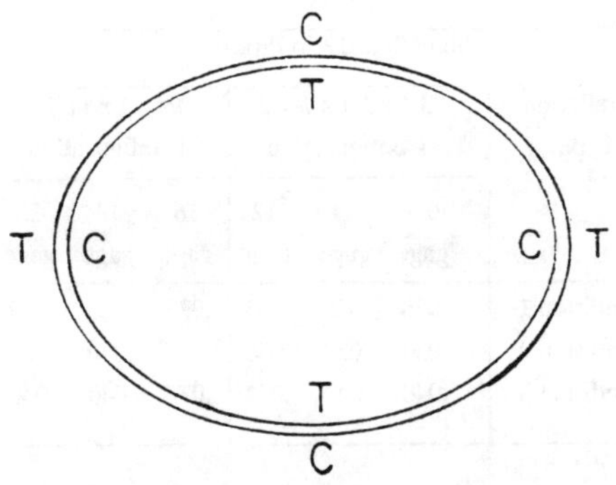

Figure 5. Bending Stresses in a Deflected Pipe

The handling and installation stiffness, relating to the moment of inertia of the pipe wall (I) is also conservatively evaluated using the net effective properties. Since deflecting a pipe requires tension in the flat elements in at least two quadrants (see Figure 5), the stiffness in these areas is represented by the full geometric moment of inertia (I). Thus, the pipe actually provides a variable stiffness around its periphery with the net effective moment of inertia representing the minimum property.

3. FLEXIBILITY LIMITS

The flexibility factor (FF) concept, represented by Equation 2 has long been used for the design of flexible metal, and more recently some plastic pipes.

$$FF = \frac{D^2}{EI} \quad \text{.......................................(2)}$$

Where:

D = Pipe diameter (in)

E = Modulus of Elasticity (psi)

I = Moment of Inertia (in⁴/in)

Flexibility factor limits traditionally have been set on the basis of experience. The goal is to ensure the pipe can be installed in the construction conditions and with the type of backfill materials anticipated. The stiffness and bending strength of the pipe must be compatible with backfill compaction and other installation loads.

Backfill and compaction loads are dependent on the type of installation -- embankment or trench -- and the compactibility of the backfill material. Trench installations typically limit compaction loads by effectively limiting the size and type of compaction equipment that can be used beside and immediately over the pipe.

The compactive effort required to bring the backfill material to 90% Standard Proctor density is dependent on several variables including material nature, gradation, moisture content, lift thickness, etc. However, the primary variable is the nature of the backfill itself. Clean, well graded granular materials and crushed rock compact with the least effort, while clayey or silty sands require more effort, resulting in more loading imposed on the pipe. Properly set flexibility factor limits must recognize the range of these compaction loads and their effect on the flexible pipe.

To develop flexibility factor limits for Spiral Rib pipes, the field performance experiences of several manufacturers were correlated. These included installation experience across a broad range of gages, diameters, backfill materials, and installation types.

The experience-based flexibility factor limits for Spiral Ribbed pipes are provided in Table 3. These limits vary with pipe material, rib configuration installation type and backfill compactability.

To simplify the flexibility factor matrix in Table 3, an empirical formula for the FF limit, Equation 3 was developed:

$$FF \leq CI^{1/3} \quad \text{.........................(3)}$$

Table 2 - Full vs. Net Effective Section Properties   (Using Table 1, Net Effective Values as a Comparison)

Steel Spiral Rib Pipes				
3/4 x 3/4 x 7-1/2 Configuration				
Full Geometric Properties			Net Effective Properties (As a % of Full)	
Thickness (in)	A (in^2/Ft)	$I \times 10^{-3}$ (in^4/in)	A (%)	I (%)
.064	.840	4.276	60.8	83.9
.079	1.048	5.367	66.2	88.2
.109	1.463	7.587	81.5	94.3
3/4 x 1 x 11-1/2 Configuration				
Full Geometric Properties			Net Effective Properties (As a % of Full)	
Thickness (in)	A (in^2/Ft)	$I \times 10^{-3}$ (in^4/in)	A (%)	I (%)
.064	.828	5.944	45.1	77.0
.079	1.034	7.462	50.6	81.5
.109	1.444	10.550	61.1	87.8

Aluminum Spiral Rib Pipes				
3/4 x 3/4 x 7-1/2 Configuration				
Full Geometric Properties			Net Effective Properties (As a % of Full)	
Thickness (in)	A (in^2/Ft)	$I \times 10^{-3}$ (in^4/in)	A (%)	I (%)
.060	.848	4.290	49.5	74.6
.075	1.051	5.389	54.4	79.0
.105	1.468	7.617	62.8	84.0
.135	1.882	9.884	69.2	88.4
3/4 x 1 x 11-1/2 Configuration				
Full Geometric Properties			Net Effective Properties (As a % of Full)	
Thickness (in)	A (in^2/Ft)	$I \times 10^{-3}$ (in^4/in)	A (%)	I (%)
.060	.831	5.965	37.5	68.3
.075	1.038	7.493	41.1	72.7
.105	1.450	10.591	48.0	79.2
.135	1.861	13.746	54.2	83.5

Where:

FF = Flexibility Factor Limit

C = A constant depending on:
Pipe Material
Rib Configuration
Backfill Type
Installation Condition

I = Pipe Wall Moment of Inertia (in^4/in)

The constant C is an accumulative result of the factors shown below Equation 3. While the form may appear unconventional, it simply provides an alternative to the matrix (Table 3).

The variations in flexibility factor limits are not significantly different than for conventional metal pipes. First, flexibility factors vary with material (aluminum or steel) and rib (corrugation) depth. These factors are necessary to account for both material stiffness (E) and bending strength differences (F_y) of the steel and aluminum materials.

Rib depth becomes a factor as is corrugation depth in conventional corrugated pipes because of the variable relationship of the geometric stiffness (I) provided versus

Table 3 - Flexibility Factor Limits For Spiral Rib Pipes

Steel Spiral Rib Pipes						
Installation Type	3/4 x 3/4 x 7-1/2 Configuration			3/4 x 1 x 11-1/2 Configuration		
	16 gage	14 gage	12 gage	16 gage	14 gage	12 gage
Condition 1	.026	.029	.033	.023	.025	.029
Condition 2	.030	.033	.038	.027	.030	.034
Condition 3	.040	.044	.051	.037	.040	.046

Aluminum Spiral Rib Pipes								
Installation Type	3/4 x 3/4 x 7 1/2 Configuration				3/4 x 1 x 11-1/2 Configuration			
	Thicknesses							
	.060	.075	.105	.135	.060	.075	.105	.135
Condition 1	.037	.040	.046	.051	.028	.031	.035	.039
Condition 2	.044	.049	.056	.062	.034	.039	.044	.048
Condition 3	.063	.070	.080	.088	.050	.055	.063	.070

Table 4 - Flexibility Factor Limits For
Conventional Corrugated Metal Pipes

Corrugation	Installation Type	Steel	Aluminum	Source
2-2/3 x 1/2	Projection Burial	.0433	.095	AASHTO[2]
2-2/3 x 1/2	Trench Burial	.060	---	AISI[3]
3 x 1	Projection Burial	.033	.060	AASHTO[2]
3 x 1	Trench Burial	.060	---	AISI[3]

bending strength (section modulus). Increasing the depth of the wall profile provides a substantial increase in stiffness (I) with a correspondingly smaller increase in bending strength or section modulus. During installation, both the pipe's stiffness and ability to resist bending strains are significant factors. A brief review of established flexibility factor limits for conventional corrugated metal pipes demonstrates the need to reduce flexibility factor limits for deeper corrugations to maintain a balance of stiffness (I) and bending strength (S) properties (see Table 4).

Finally, to account for the variation in both basic construction conditions and backfill materials used in the pipe zone, experience dictated three categories to represent installation conditions. These are:

- Condition 1 - Standard embankment burial with standard backfill materials
  (AASHTO M 145; A-1, A-2 and A-3)
- Condition 2 - Trench burial with standard backfill materials
  (AASHTO M145; A-1, A-2 and A-3)
- Condition 3 - Trench burial with select, easily compacted backfill materials
  (AASHTO M145; A-1, A-2-4 and A-2-5)

While a greater number of conditions could be developed, recognizing further variations in backfill materials, limits on compaction equipment, lift thickness, etc., these three provide a reasonable recognition of field conditions while providing an understandable and controllable breakdown.

A visual comparison of flexibility factor limits for Spiral Rib pipes and conventional corrugated metal pipes (Tables 3 and 4) demonstrates the lower limits established for Spiral Rib pipes when installation conditions are accounted for. Spiral Rib pipes, with their deep rib configuration are typically less flexible than conventional corrugated metal pipes.

4. CONCLUSIONS

Spiral Rib metal pipes, while utilizing a novel ribbed pipe wall configuration to maintain essentially "smooth wall" flow conditions, are designed on the same basis as conventional corrugated metal pipes. Design is facilitated through the use of:

- Net effective section properties that ensure the structural performance ofthe pipe wall.
- Experience based flexibility factor limits that account for the different rib configurations and materials, as well as installation and backfill conditions.
- Three easily recognized installation and backfill conditions to facilitate design and control.

REFERENCES

1. Cold Formed Steel Design Manual - American Iron and Steel Institute, Washington, DC; 1980
2. Standard Specification for Highway Bridges - American Association of State Highway and Transportation Officials; Washington, DC; 1983
3. Modern Sewer Design - American Iron and Steel Institute, Washington, DC; 1980.

Structural Performance of Flexible Pipes, Sargand, Mitchell & Hurd (eds) © 1990 Balkema, Rotterdam. ISBN 90 6191 165 6

Construction and inspection report on smooth lined corrugated polyethylene pipe

L.John Fleckenstein & David L.Allen
Kentucky Transportation Center, University of Kentucky, Lexington, Ky., USA

ABSTRACT: This paper documents the installation and performance of corrugated polyethylene pipe (CPEP) installed during five construction projects in Kentucky. The pipe was manufactured by Advanced Drainage Systems, Inc., and is designated as ADS N-12. The pipe is composed of high-density polyethylene (HDPE) with a corrugated exterior and a smooth interior.

The CPEP requires less equipment and fewer personnel than steel or concrete pipe for installation. Extreme care should be taken during backfilling around the pipe. Due to the fact the CPEP is lightweight, the pipe has a tendency to rise or drift during backfilling. To eliminate this, the contractors bed each side equally to approximately 1/2 to 3/4 of the pipe height before compacting (depends greatly on the backfill material). Cuts or tears were observed inside the CPEP pipe (5 total). The tears probably occurred due to improper backfilling and unequal loading of the pipe wall. Select backfill should be provided up to one-foot over the crown of the pipe to protect the pipe against backfill damage. Care should be taken not to damage the plastic pipe during transportation.

1 INTRODUCTION

This report documents the installation and performance of corrugated polyethylene pipe (CPEP) installed during construction of South Forbes Road in Fayette County, KY 54 in Daviess County, US 62 in McCracken County, US 68/KY 80 in Warren County and Nicholasville Road in Fayette County. The pipe was manufactured by Advanced Drainage Systems, Inc., and is designated as ADS N-12. The pipe is composed of high-density polyethylene (HDPE) with a corrugated exterior for increased strength and a smooth interior to provide maximum flow capacity.

The purpose of the study was to evaluate the performance of the pipe during construction and after placement.

2 FORBES ROAD INSTALLATION

In November 1987, the first section of pipe installed in Kentucky was along Forbes Road in Fayette County. The pipe was installed in two locations. The first was 15 inches in diameter and was installed 28 feet right of Station 6+54. The 15-inch CPEP was used as an entrance pipe (very seldom loaded). The entrance pipe was backfilled with approximately one foot of material.The second location started at a storm sewer inlet and ran 240 feet north to a manhole; this pipe is used as a 15-inch culvert (Figure 1). The pipe was backfilled with No. 9 stone to approximately one foot above the top of the pipe. The remainder of the trench was backfilled to grade elevation with excavated material. The maximum fill height was approximately 6 feet.

The pipe was visually inspected on November 19, 1987, and September 21, 1989. Random measurements were taken of the internal diameter of both pipes. Measurements indicated there had been little to no distortion of the pipe since construction. The pipes also were inspected for chemical or physical deterioration or defects. The inlet, outlet, and manholes were examined and photographed. There were no signs of deterioration or weakening of the pipe.

3 KY 54, DAVIESS COUNTY

3.1 Construction Inspection

During the relocation of KY 54, in Daviess County, approximately 468 feet of 15-inch pipe, and 592 feet of 18-inch pipe were installed.

The project begins at Station 93+81.17 and ends at Station 152+00. The first section of pipe was placed May 18, 1988, at Station 137+88. Approximately 104 feet of 18-inch pipe were placed at that location. The trench was backfilled with a coarse and clean sand approximately midway up the pipe and compacted with a vibratory tamper. The remainder of the trench was backfilled with excavated material (Figure 2). The excavated material was alluvial by nature (sandy silt).

Prior to installation, it was observed that a number of the pipe sections had been cut during construction. It appeared that the sections had been damaged during transport with a backhoe. The damaged sections were returned to the manufacturer for inspection (Figure 3).

3.2 Monitoring Points

Two sets of monitoring points were placed in the pipe at Station 137+88 prior to installation. The monitoring points were placed at 41 feet and 61 feet from the outlet. The points were monitored three times during 1988 and on September 18, 1989. Since installation, the interior of the pipe at 61 feet had deflected approximately 0.42 inch, and approximately 0.25 inch at 41 feet (Figures 4 and 5). The fill height above the 61 foot monitoring point was 6 feet, and the fill height at the 41-foot monitoring point was 8 feet.

Monitoring points were also placed at Station 126+00. Since installation, the interior of the 18-inch pipe has deflected 0.34 inch (Figure 6). Eight additional monitoring points were placed in the pipes prior to installation. Due to water and mud, the monitoring points

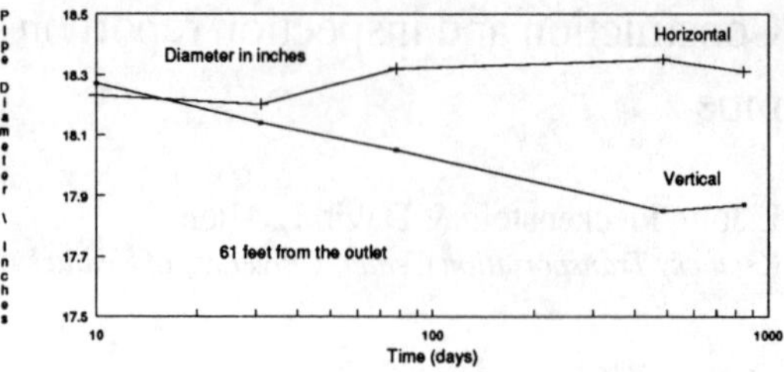

Figure No. 4: 18-Inch CPEP   KY 54, Daviess County, Station 137+88

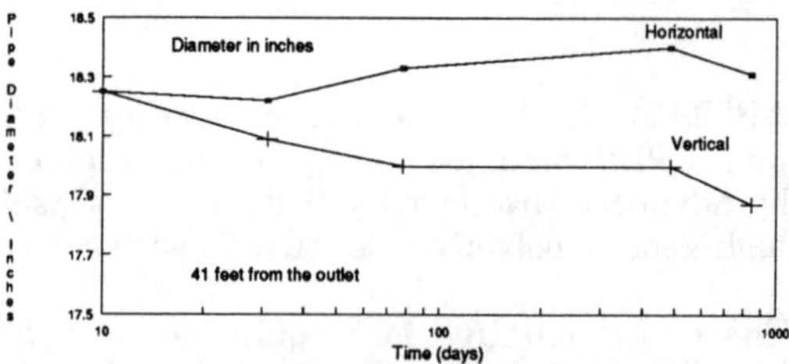

Figure No. 5: 18-Inch CPEP   KY 54, Daviess County, Station 137+88

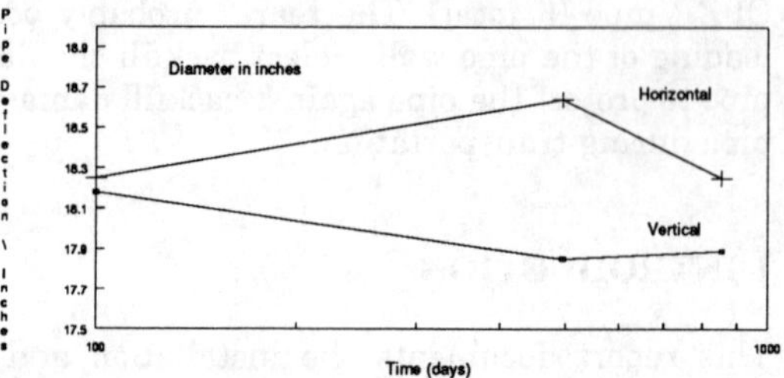

Figure No. 6: 18-Inch CPEP   Ky 54, Daviess County, Station 126+00

could not be accessed during the last inspection.

3.3 Visual Inspection

All pipes, including the 18-inch and 15-inch pipe were inspected and photographed. All appeared to be symmetrical. On June 18, 1988 during the inspection of the 104-foot, 18-inch pipe at Station 137+88, a rip or cut was noted in the smooth interior lining of the pipe. The rip follows one of the corrugations and is approximately seven inches in length (Figure 7). The rip was closely inspected again on September 18, 1989. The rip appeared to be approximately the same length and had bulged approximately one-half inch.

The interior of a 216-foot, 18-inch pipe was inspected in detail on September 19, 1989, from Station 126+00 to 128+20. Each 20-foot section was photographed and visually inspected. The pipe appeared to be clean and symmetrical. A four-inch gap was discovered between the first and second section south of Station 128+20 (Figure 8). It was apparent that the sections were not completely butted together at the time of installation. All other joints were inspected and their conditions were recorded. The next largest separation was one inch. A one-half inch separation at joints was fairly common throughout the pipe section.

4 US 62, McCRACKEN COUNTY

During 1989-1990, approximately 652 feet of N-12 corrugated polyethylene pipe was installed during a reconstruction of US 62 from two lane to four lane. As shown in Table 1, the sizes of pipe installed were 15, 18, 24, and 30 inches. The majority of the pipe was being used as entrance pipes for local residences. On September 18, 1989, all entrance and culvert pipes were visually inspected and photographed. Several of the pipes showed signs of deflection. It appears the pipes had deformed during backfill operations. Appendix A is an inspection log for each pipe.

On March 13, 1990, all pipes were visually inspected, photographed, and deflection measurements were taken. Contained in Appendix B is an inspection log of each pipe. Through further observation it appears that culvert pipes (cross drains) are functioning well with little notable deflection. The pipe used as entrance pipe appears to be deflecting considerably more than pipe use for culverts (Figure 9). It appears that the entrance pipes are deflecting due to shallow fill heights, weak (alluvial) backfill material, and traffic loading. A 9-inch deep, 1.5-foot long compression failure was observed in an 18-inch pipe located under a concrete driveway (Figure 10).

5 US 68/KY 80, WARREN COUNTY

In 1989, a contract was awarded for the widening of US 68/KY 80. Included in this contract was the installation of pipe. As shown in Table 1, approximately 7,080 feet of pipe will be installed. The pipe will be used predominately as storm drains and cross drains. As of May 15, approximately half of the pipe had been installed. Monitoring points were marked inside the pipes on November 28, 1989 while the pipes were stockpiled. A puncture was observed inside one of the 36 inch pipes. The puncture passed through the outside corrugations and through the smooth interior lining.

A construction and monitoring inspection was performed on December 12, 1989, January 11, and February 21, 1990. The pipe was being backfilled with No. 11 stone to an elevation approximately 3/4 up the height of the pipe. The material was being compacted with vibratory compactors. On December 12, 1989, the remaining trench was backfilled with the excavated material which consisted of frozen red to brown clay and rock (Figure 11). According to Section 612.05 of "The Kentucky Standard Specifications For Road And Bridge Construction", the fill material should be free of rocks larger than 3 inches and is not to contain frozen clods of soil. A large section of 36-inch pipe was visually inspected and deflection measurements were obtained (Table 2). As shown in Table 2, there is only moderate pipe deflection.

On April 18, 1990, deflection measurements were obtained on other sections of 36-inch and 24-inch pipe (Table 2). During the inspection, a tear approximately 6 inches long was observed in a 24-inch pipe. The pipe had deflected 3 inches, approximately 45 degrees off vertical. The tear had occurred in one of the inside spiral corrugations. It appears to have been caused by a rock in the backfill. Two tears were observed inside the corrugated polyethylene pipe in Warren County. It appears the tears are occurring where the sections of plastic are wrapped together to form the pipe. The tears may be occurring due to improper backfilling and unequal loading of the pipe. It appears that large rocks or large clods are becoming lodged against the pipe during backfilling operations.

6 NICHOLASVILLE ROAD, FAYETTE CO.

Approximately one mile of CPEP is being installed during a widening project on

Nicholasville Road in Fayette County (Figure 14). As of May 17, 1990, approximately one half of the pipe has been installed. The pipe is being used as cross drains and storm drains. The trench around the pipe is being backfilled with No. 9 stone to an elevation approximately 1 foot above the pipe. The remainder of the trench is being backfilled with excavated material (red/brown clay). There were no visual signs of deflection during a visual inspection conducted on May 17, 1990. The pipe appeared to be in excellent condition.

7 DISCUSSION

The CPEP requires less equipment and fewer personnel than metal or concrete pipe for installation. Extreme care should be taken during backfilling around the pipe. On US 62, KY 54, and US 68/KY 80, the pipes were not completely covered with bedding material. Cuts or tears were observed inside the N-12 pipe (5 total). It appears that the tears are occurring where the sections of plastic are wrapped together to form the pipe. The rips are probably occurring due to improper backfilling and unequal loading of the pipe wall. On Forbes Road and Nicholasville Road, the pipes were completely covered with one foot of sand or crushed stone before the remainder of the trench was backfilled with excavated material. This one foot of cover helps to protect the pipe against backfill damage. On several of the installations, it was apparent the ends of the pipes at the couplings are rarely butted completely together. This permits material to be deposited in this area. Care should be taken not to damage the plastic pipe during transportation.

8 CONCLUSION

Extreme care should be taken during backfilling around the pipe. Due to the fact the CPEP is lightweight, the pipe has a tendency to rise or drift during backfilling. To eliminate this, the contractors bed each side equally to approximately 1/2 to 3/4 of the pipe height before compacting (depends greatly on the backfill material).

Select backfill should be provided up to one foot over the crown of the culvert.

Table 1.

Location	Diameter (inches)	Quantity (feet)
Fayette (Forbes Rd.)	15	255
Daviess (KY 54)	15	480
Daviess	18	592
McCracken (US 62)	15	272
	18	112
	24	144
	30	124
Warren (US 68/KY 80)	15	4,068
Warren	18	1,132
Warren	21	624
Warren	24	756
Warren	30	64
Warren	36	436
Fayette (Nicholasville Rd.)	12	204
Fayette	15	732
Fayette	18	2,400
Fayette	24	1,460
Fayette	30	20
Fayette	36	320
Total		14,195

Table 2. Deflection Measurements, US 68/KY 80, Warren County

Pipe ID	Initial Measurement (inches)	2-21-90	4-18-90
2V (Vertical)	36.25	35.75	
1V (Horizontal)	35.87	36.31	
1N (Vertical)	35.87	35.81	
2N (Horizontal)	36.0	35.81	
2Q (Vertical)	35.5	35.81	
1Q (Horizontal)	35.87	36.09	
2S (Vertical)	35.68	35.68	
1S (Horizontal)	36.06	35.87	
1R (Vertical)	36.12	36.12	
2R (Horizontal)	35.93	35.75	
2U (Vertical)	36.12	36.25	
1U (Horizontal)	36.18	35.87	
2T (Vertical)	35.87	35.25	
1T (Horizontal)	36.0	36.37	
1X (Vertical)	35.5	35.56	
2X (Horizontal)	36.0	35.78	
2F (Vertical)	36.0		35.75
1F (Horizontal)	35.81		35.87
2K (45 degrees)	35.87		35.75
1K (45 degrees)	35.93		35.75
2KA (Vertical)	(none)		35.18
1KA (Horizontal)	(none)		36.87
1A (Vertical)	36.18		36.0
2A (Horizontal)	35.75		35.37
1H (Vertical)	24.12		24.0
2H (Horizontal)	24.43		24.5
2J (45 degrees)	24.0		24.18
1J (45 degrees)	24.0		23.75
2P (Vertical)	23.93		23.81
1P (Horizontal)	24.0		23.87
2D (Vertical)	23.75		23.68
1D (Horizontal)	23.56		23.75
2M (45 degrees)	23.87		23.81
1M (45 degrees)	24.06		24.12
1B (Vertical)	24.12		23.87
2B (Horizontal)	24.0		24.25
1O (45 degrees)	23.68		23.75
2O (Buried)	23.31		

Figure 1. Backfill around 15-inch culvert
(Forbes Road).

Figure 2. Compacting sand backfill around
18-inch culvert (KY 54).

Figure 3. Damaged 18-inch N-12 pipe (KY 54).

Figure 7. Rip or cut inside an 18-inch
pipe (KY 54).

Figure 8. Four-inch separation between the two connecting
pipe ends at the coupling (KY 54).

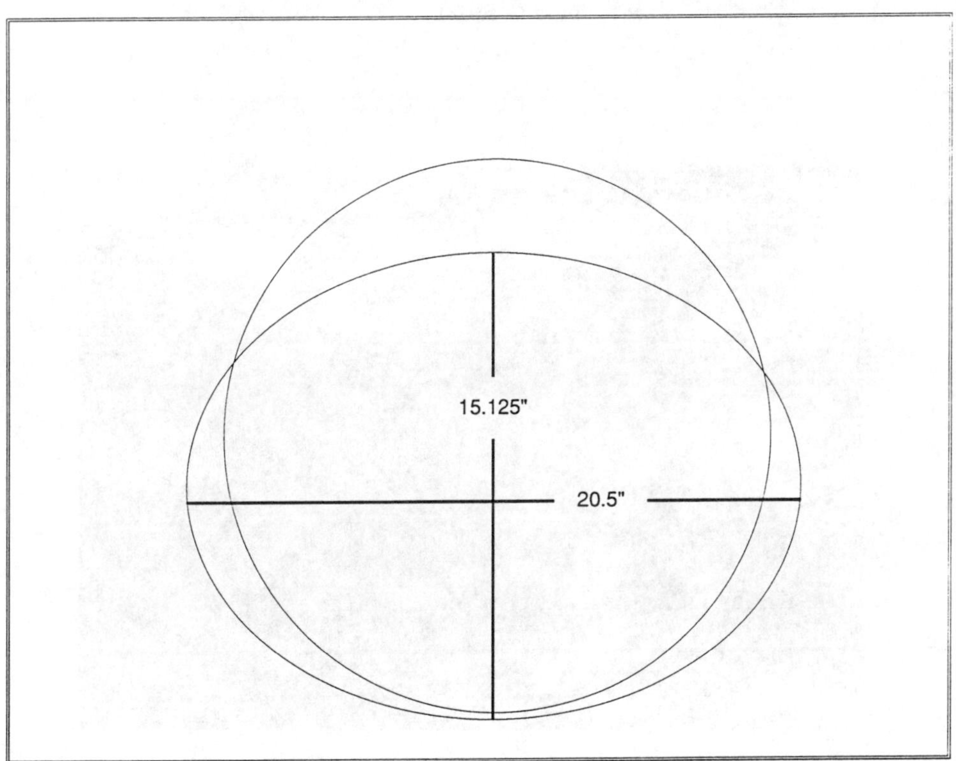

Figure 9: 18-Inch N-12 pipe, US 62, McCracken Co.
Station 180+00, Entrance Pipe

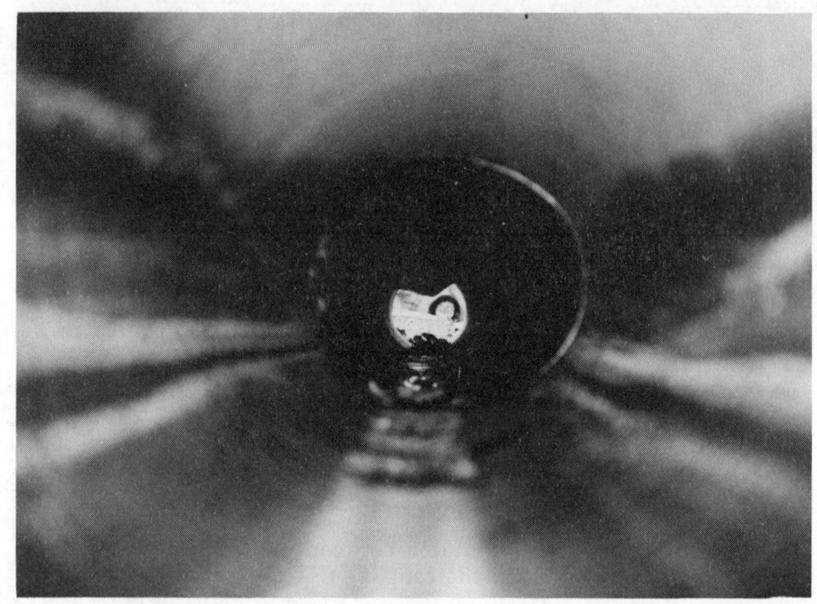

Figure 10. Failure within 18-inch entrance
pipe (US 62).

Figure 11. Backfilling of 15-inch culvert pipe (US 68/KY 80).
Soil appears to be clodded and frozen.

APPENDIX A
US 62 N-12 PIPE INSPECTION
SEPTEMBER 18, 1989

SITE   DESCRIPTION

No. 1   18-inch entrance pipe slightly irregular.

No. 2   24-inch entrance pipe (Station 174+00). Approximately 2.3 inches of deflection at the pipe connection. Some settlement observed in the gravel drive above the pipe. Approximately 1.5 feet of cover at the south end of the pipe, south side of the gravel drive. The pipe sections are not completely connected.

No. 3   18-inch entrance pipe (Station 167+50). The pipe has approximately 3 inches of cover at 6 feet north of the south end. Approximately 2 inches of compression noticeable at this point. Endloader tracks are also present on the surface at this point.

No. 4   18-inch entrance pipe (Station 161+50). Slight bow present at the joint in the pipe. Overall installation looks good. Asphalt surface has been placed within 1 to 2 inches of the top of the pipe.

No. 5   18-inch entrance pipe (Station 151+00-Highland Road). The installation is in good condition.

No. 6   24-inch entrance pipe (Station 138+50). Approximately 2 inches of compression on the south end. Approximately 4 to 4.5 feet of fill. Pipe does not appear to be bedded on sand or crushed stone.

No. 7   24-inch culvert pipe (Station 134+50). Good installation.

No. 8   30-inch culvert pipe (Station 124+50). Slight vertical deflection approximately 25 inches from the east end. Fairly good shape.

No. 9   15-inch entrance pipe (Station 121+00). Truck entrance for Turner Dairy. Half full of soil. Not much deflection.

No. 10   15-inch entrance pipe (Station 111+00). Half full of soil. Little to no deflection.

SITE	DESCRIPTION	DEFLECTION READING	
		(Vertical)	(Horizontal)
No. 1	18-inch entrance pipe (Station 180+00).	15.12"	20.5"
No. 2	24-inch entrance pipe (Station 174+00). Approximately 3-4 inches of deflection Not able to measure due to sediment. New concrete drive.		
No. 3	18-inch entrance pipe to field (Station 167+50). Backfill has been washed away.	16.87"	19.87"
No. 4	18-inch entrance pipe (Station 161+50).	17.0"	18.75"
No. 5	18-inch entrance pipe (Station 151+00-Highland Road). The installation is in good condition.		
No. 6	24-inch entrance pipe (Station 138+50).	22.83"	25.25"
No. 7	24-inch culvert pipe (Station 134+50).	22.25"	24.43"
No. 8	30-inch culvert pipe (Station 124+50).	29.5"	30.16"
No. 9	15-inch entrance pipe (Station 121+00). New concrete drive has been recently installed. Approximately 2 inches of defection in the pipe.		
No. 10	15-inch entrance pipe to field (Station 111+00). Fence blocking access. Pipe appears to be in good shape.		
No. 11	18-inch entrance pipe to residence with new concrete drive. Indentation in pipe approximately 1.5 feet in length.	9.25"	21.75"

Structural Performance of Flexible Pipes, Sargand, Mitchell & Hurd (eds) © 1990 Balkema, Rotterdam. ISBN 90 6191 165 6

Nine year performance review of a 24″ diameter culvert in Ohio

James B.Goddard
Advanced Drainage Systems, Inc., Columbus, Ohio, USA

ABSTRACT: The first 24" diameter corrugated polyethylene pipe crossdrain installed by a state highway department in the United States was installed by the Ohio Department of Transportation in 1981. The site selected is extremely hostile in terms of acidity and abrasion. The pipe condition has been monitored over the past 9 years and its condition is reported. Pipe deflection was measured four times, and results compared with theoretical values generated using CANDE and controlled test results. Correlation of those values are very good. Particular attention was paid to invert condition, specifically the effects of abrasion.

1 INTRODUCTION

In early 1981, 18" and 24" diameter corrugated polyethylene pipe was manufactured in the United States, primarily for culvert applications. On September 9, 1981, the Ohio D.O.T. installed 40 linear feet of 24" diameter corrugated pipe as a cross drain under State Route 145 in Noble County, 3.59 miles north of the county line (Figure 1). The site had been a persistent maintenance problem because a highly acidic and abrasive flow had caused total failure (i.e. collapse) or complete invert deterioration of several pipes in from two to five years. The pipe replaced in 1981 was a Nexon coated corrugated steel pipe culvert (Figure 2).

Based on the condition of the samples of polymeric coating removed from the site, it appeared that the coating was abraded by the sand and gravel carried through the pipe from an abandoned strip mine approximately 1/8th mile upstream. As the zinc coating under the polymeric coating was exposed, it corroded rapidly, undercutting and lifting the film from the steel. The exposed zinc and steel were rapidly removed by the acidic mine runoff, which has a pH of from 2.5 to 4.0, depending on flow volume.

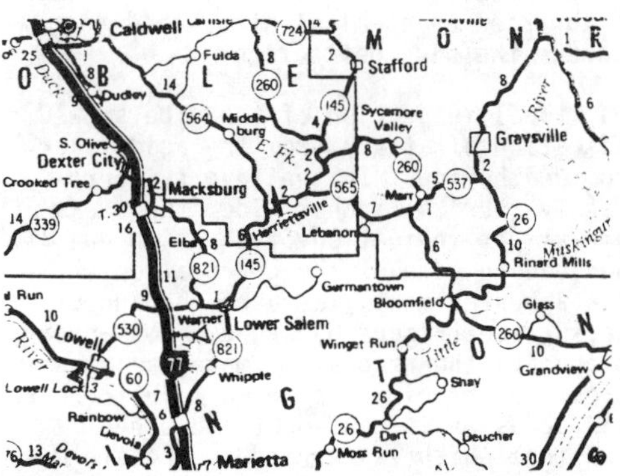

Figure 1. Location of 24" CPEP culvert.

Figure 2. Failed corrugated steel pipe.

Figure 3. 24" diameter CPEP in 2-20' lengths.

Figure 4. First 20' section in trench.

Figure 5. Completed backfill.

2 INSTALLATION

The installation of the corrugated polyethylene pipe was done by a state maintenance crew from District 10, Marietta, Ohio, under the direction of Mr. Don Johnson and Mr. Dale Hockenberry. The pipe was delivered to the jobsite in two 20' lengths on a truck with a 12' bed (Figure 3). A 30" wide trench was dug in a soil composed of rock and clay. A bed of 4 inches of bank run gravel was placed and leveled. The pipe was then placed in the trench (Figure 4) and backfilled to just over the top of the pipe and compacted in the same manner (Figure 5). Then about 6 inches of cold patch was placed over the soil, compacted and leveled.

To maintain traffic, only one half of the road was opened at a time. The above process was done on each side, with traffic maintained over the first section (downstream), while the second section was installed. The two sections of trench were not perfectly aligned, with an approximate 7° horizontal offset at the joint.

Traffic along State Route 145 consists of local residents' cars and trucks, coal trucks, and oil tank trucks. This traffic was maintained throughout the installation, with the roadway reduced to one lane for a total of five hours.

3 INSPECTIONS

On November 15, 1982, Ohio D.O.T. and the manufacturer's personnel returned to inspect the installation. The intent of the trip was to inspect the pipe for signs of abrasion or chemical attack and to measure the pipe deflection.

A visual inspection of the pipe showed no indication of attack from the acidity or from weathering. In the invert, approximately four (4) inches wide, there was some slight indication of abrasion on the leading edge of the corrugation. This appeared mostly as a roughening of the surface (Figure 6) and did not appear to involve the removal of material; therefore, the abrasion was not a major concern. This abrasion is caused by a crushed sandstone which is carried by the storm flows from a hollow fill which has not been reclaimed.

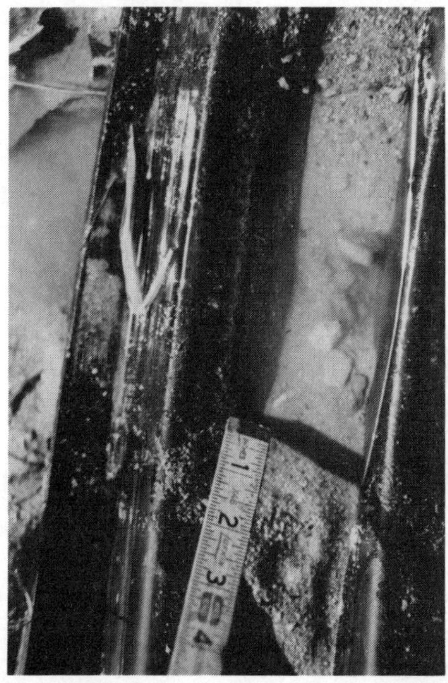

Figure 6. Pipe invert at the outlet end.
Note the size and angularity of the stones
and sand carried through the pipe. Note
slight roughening of the leading edge of
the corrugation crown.

Deflection measurements were taken using
the probe designed for the "Effect of
Heavy Loads on Buried Corrugated Polyethylene Pipe" study (T.R.R. 903) with
deflections measured every foot of the pipe
length. The deflection readings are shown
in Table 1. The deflection reading at the
pipe joint is affected by the bend put in
the pipe at installation and does not
appear to have been caused by loadings.
Regardless of the cause, 7.3% deflection is
within the design limits of corrugated
polyethylene pipe. The deflection at the
point 9' from the outlet end is indicative
of that caused by repeated vehicle loads.
At that point, there is a depression in the
roadway shoulder caused by repeated vehicles running off the roadway. Cover at that
point is less than 15".

The average deflection for the pipe that
is under load is 1.9%, which is substantially below all accepted flexible pipe
design standards.

The site was inspected visually again in
1983, and no change in condition had
occurred.

In May of 1985, the site was again
visited as part of John Hurd's inspection
of 172 corrugated polyethylene pipe culverts (T.R.R. 1087). At that time, deflection measurements were again taken. These
values had not changed from those found in
November of 1982. Site conditions had
changed in that Route 145 had been repaved
since the last inspection from the county
line to about one mile past the culvert.
There was no indication of pavement cracking or settlement over the pipe at this
time.

A pile of sediment consisting of small
rocks, coal, and sand on the bank of the
drainage course 35 foot long, 15 foot
wide, and about 12" deep indicated the
force of the storm flows through the
culvert and the abrasive nature of those
flows. Despite this, the pipe was in
good condition with little change in the
appearance of the pipe invert from the
1982 inspection.

In July of 1988, the site was inspected
again. Since the last inspection (1985),
the upstream ditch had been realigned so
that the flow ran in a nearly straight
path over large stone riprap to the pipe
inlet. It appeared this has reduced the
headwater level. The pipe itself appeared
unchanged, with deflection still averaging
1.9% for the length under load. There
was no cracking or settlement in the
pavement above the pipe. The exposed
pipe end showed no signs of damage and
were still shiny black, indicating little
oxidation or UV effects.

In May of 1990, the site was inspected
again. Again, the pipe appeared unchanged,
with no change in deflection. There was
no change in the pavement above the pipe.
The realignment of the upstream ditch
appeared to have caused an increase in
velocity through the pipe causing a deep-
ening and widening of the ditch at the pipe
outlet. This did not appear to have
effected the pipe invert.

Comparing measured deflection values with
test results from Dr. Reynold Watkins'
work (Ref. 4) and Dr. Mike Katona's (Ref.
2) theoretical values demonstrates an
excellent correlation. From Dr. Watkin's
deflection curves (Figure 9), the predicted
deflection of a 24" diameter corrugated
polyethylene pipe would be 1.5% with 95%
standard AASHTO density backfill and 2.5%
with 90% standard AASHTO density backfill
and 15 inches of cover. Dr. Katona's
predicted deflection is 2.3% at 95% compac-

Table 1. Ohio D.O.T. 24" test installation installed 9/9/81. Tested for deflection 11/15/82.

Feet From Outlet	Deflection (inches)	Deflection (%)
1	0	0
2	0	0
3	0	0
4	0	0
5	0	0
6	0	0
7	0	0
8	0	0
9	0.5	2
10	0.375	1.5
11	0.125	0.5
12	0.125	0.5
13	0.0625	0.2
14	0.125	0.5
15	0.250	1.0
16	0.250	1.0
17	0.375	1.6
18	0.875	3.6
19	1.25	5.2
20 - Joint	1.75	7.3
21	1.125	4.7
22	1.0	4.2
23	0.875	3.6
24	0.75	3.1
25	0.625	2.6
26	0.75	3.1
27	0.625	2.6
28	0.375	1.5
29	0.250	1.0
30	0.125	0.5
31	0	0
32	0	0
33	0.250	1.0
34	0.50	2.0
35	0.250	1.0
36	0.250	1.0
37	0.125	0.5
38	0.125	0.5
39	0	0
40	0	0

Average 40'   1.44%
Average 30'   (Disregard Ends) 1.92%

tion and 15 inches of cover. The measured value at this site of 1.9% average deflection fits well with this data.

4 CONCLUSIONS

Based on the inspections and the nearly nine years of experience with this pipe installation, the following conclusions can be drawn:

Figure 7. Deflection probe in place at pipe outlet. Deflection readings and profile are shown in Table 1 and Figure 8.

1. There has been no increase in pipe deflection since the first inspection in 1982. Deflections appear to be largely installation caused.

2. Within the pH range present at this site, corrugated polyethylene pipe is unaffected by acid mine run-off.

3. Corrugated polyethylene pipe appears to be highly resistant to abrasion from crushed sandstone and coal at fairly high velocities.

4. Exposed pipe ends of this thickness appear unaffected by UV exposure over this time period.

5. Pavement cracking or settlement does not occur with time over a corrugated polyethylene pipe installed with reasonable care.

6. Deflection values measured at this site fit well with theoretical and controlled test values developed by others.

REFERENCES

Hurd, J.O. 1986. Field performance of corrugated polyethylene pipe culverts in Ohio. Washington, D.C.: Transportation Research Record 1087, Transportation Research Board.
Katona, M.G. 1990. Minimum cover heights for corrugated plastic pipe under vehicle loading. Washington, D.C.:

Transportation Research Board.
Katona, M.G. 1988. Allowable fill heights
 for corrugated polyethylene pipe.
 Washington, D.C.: Transportation
 Research Record 1191, Transportation
 Research Board.
Watkins, R.K.; Reeve, R.C.; and Goddard,
 J.B. 1983. Effect of heavy loads on
 buried corrugated polyethylene pipe.
 Washington, D.C.: Transportation Research
 Record 903, Transportation Research Board.

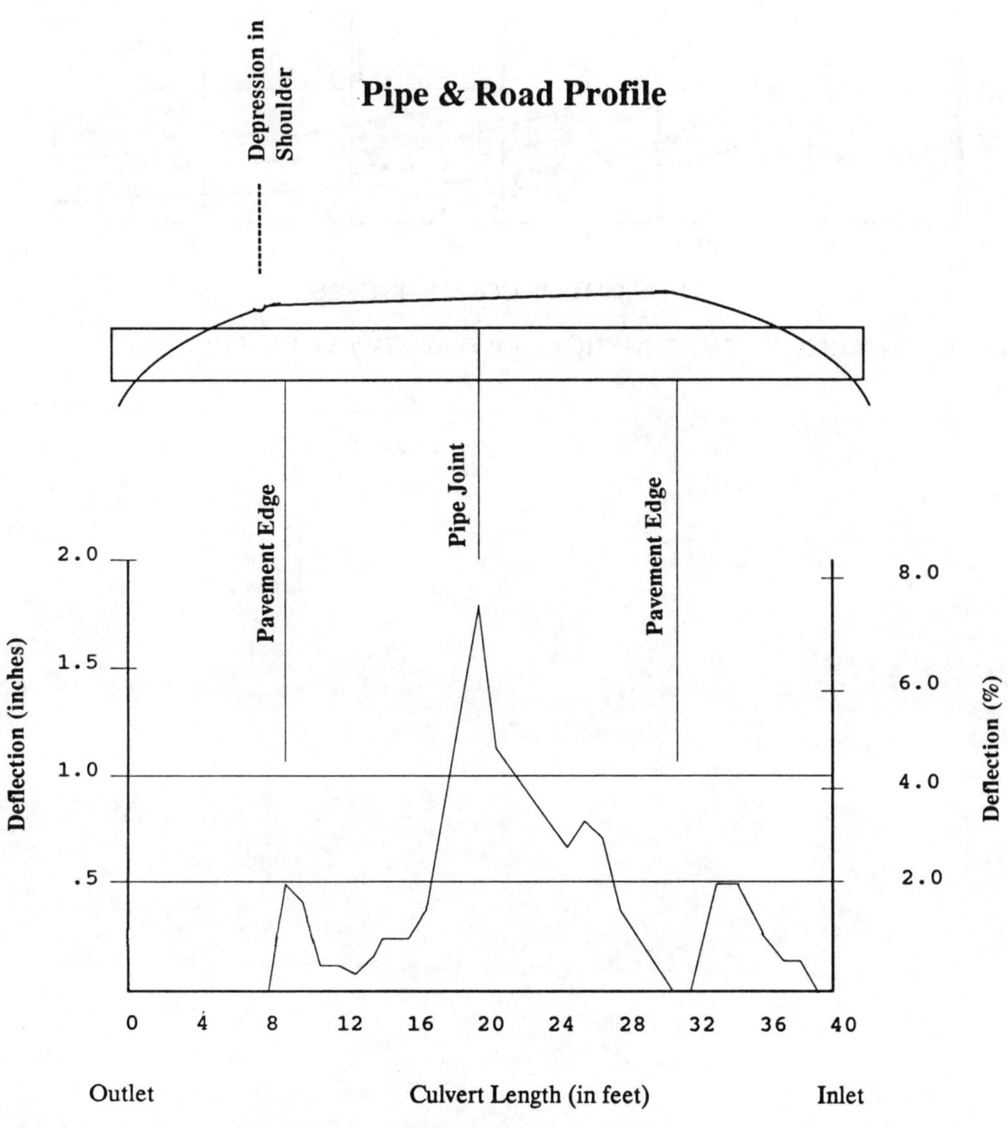

Figure 8. Pipe profile and deflection curve.

93

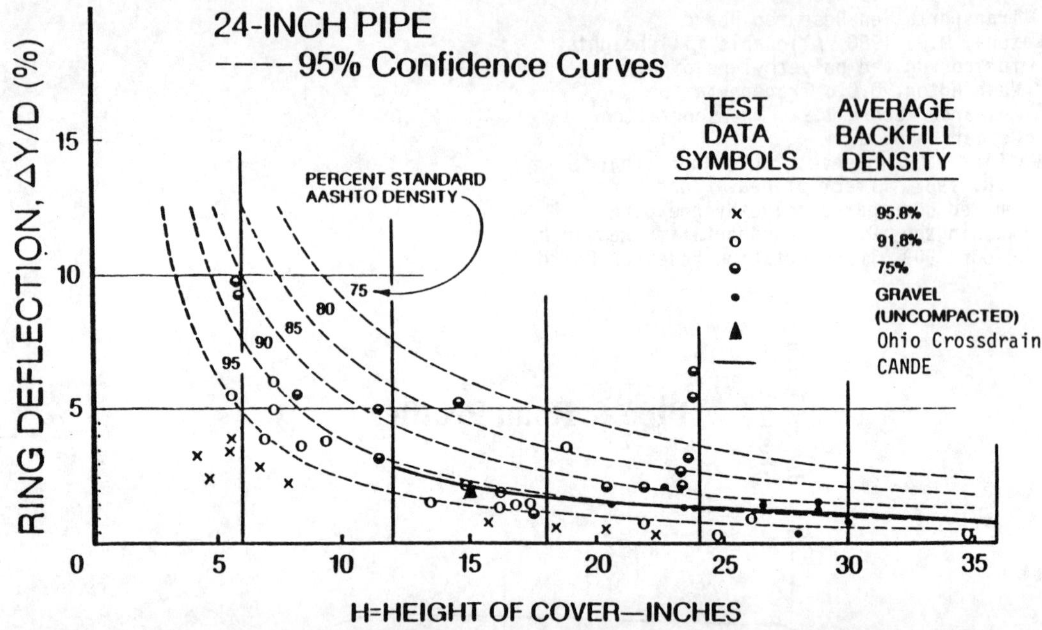

Figure 9.   COMPARISON OF TEST, THEORETICAL, AND FIELD MEASURED DEFLECTION VALUES

Structural Performance of Flexible Pipes, Sargand, Mitchell & Hurd (eds) © 1990 Balkema, Rotterdam. ISBN 90 6191 165 6

Analysis of the performance of a buried high density polyethylene pipe

Naila Hashash & Ernest T. Selig
University of Massachusetts, Amherst, Mass., USA

ABSTRACT: This paper examines the performance of a high density polyethylene pipe under high fill. The basis is a field installation where various pipe and soil parameters were monitored during construction as well as after construction. With 100 ft (30.5 m) of cover, and 722 days after pipe installation, the soil compressive strains were 1.7% in the pipe trench and 4.2% 16 ft (4.9 m) away from the pipe. The horizontal stress at springline level in the trench was 58 psi (400 kPa) and the vertical stress on the pipe crown in the trench was 22 psi (152 kPa). At the springline level, the pipe wall compressive strain was 1.4% and the bending strain was 0.3%. The vertical diameter decreased by 4.3% and the horizontal diameter increased by 0.6%. This diameter change is a result of circumferential shortening of 1.6%. Two finite element codes were used to analyze the results. To approximate the viscoelastic nature of the pipe with static analysis, a reduced pipe modulus was used. The computations compared reasonably well with the field measurements. The analysis was useful in estimating the hoop thrust in the pipe wall which could not be directly measured. Arching of earth load defined in terms of pipe springline hoop thrust reduction was 77%.

1 INTRODUCTION

Little field experience has been available for high density polyethylene (HDPE) pipes under high earth loads. To provide information on pipe performance in this application, a field test installation was carried out with a 24-in. (610-mm) inside diameter HDPE pipe placed beneath a 100-ft-high (30.5-m) embankment. This embankment is part of the extension of Interstate Route 278, North of Pittsburgh, Pennsylvania. The HDPE pipe was not required to carry any flow. It was present just for the research. The actual water flow through the embankment was carried by a 72-in. (1.8-m) diameter concrete pipe placed adjacent to the test installation.

Instrumentation was installed to measure pipe wall strains, pipe circumference change, pipe diameter change, vertical soil strain adjacent to the pipe in the free field, vertical strain of the backfill in the trench, earth pressure acting on the pipe, and temperatures of the pipe wall and air inside in the pipe. These measurements were monitored during construction, and subsequently, for 2 years following pipe installation.

A previous paper by Adams, Muindi and Selig (1989) described the installation and instrumentation, and presented results obtained during embankment construction. After briefly reviewing the field installation, this paper presents the long term performance results and interpretation of the results using finite element analysis.

2 PIPE INSTALLATION AND INSTRUMENTATION

The 576 ft (176 m) long installed pipe consists of both corrugated and smooth interior wall sections. Most of the upstream sections had a smooth interior wall. Two corrugated sections located below the center of the embankment where the fill height is maximum were instrumented. Both type pipes had a minimum stiffness of 34 psi (234 kPa) in accordance with AASHTO specifications (AASHTO, 1986a), an inside diameter of 24 in. (610 mm), and a wall thickness of 0.22 in. (5.6 mm). The pipe was manufactured in 20-ft (6.1-m) lengths.

The pipe installation conditions are shown in Fig. 1. A detailed description of the installation is presented by Adams, et al. (1988, 1989). Prior to the pipe installation, a 15-ft-high (4.6-m) embankment was constructed above the existing ground. The embankment material was silty, clayey gravel and sand. A 5-ft (1.5-m) deep and 6-ft (1.8-

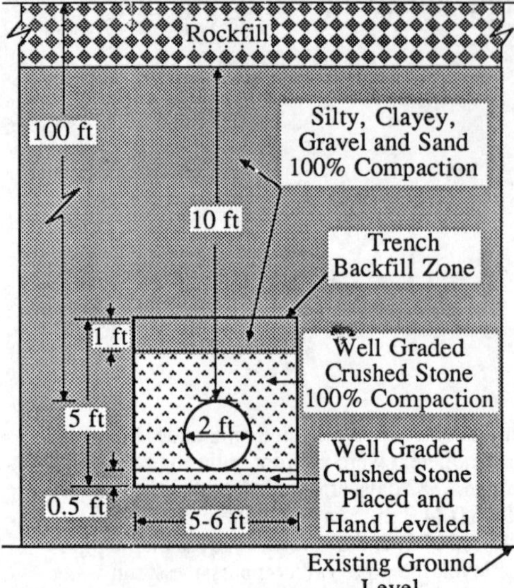

Fig. 1   Polyethylene pipe installation

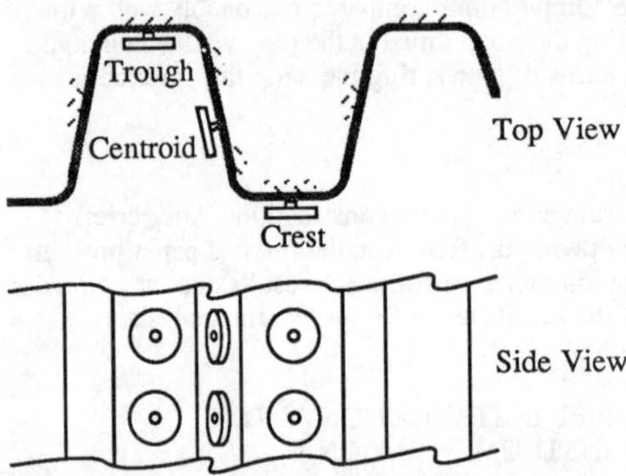

Fig. 2   Strain coils at springline level

m) wide trench for the HDPE pipe was then excavated. A 6-in. (152-mm) thick layer of structural backfill was deposited in the bottom of the trench as pipe bedding. This structural backfill material is a broadly graded crushed limestone. After the pipe was assembled in the trench, a 12-in. (305-mm) layer of backfill was placed to bring the fill up to the springline and then compacted. This was followed by 5 layers of structural backfill, 6 in. (152 mm) thick. The rest of the trench is backfilled with embankment material compacted to 100% T-99. The embankment material was then added in layers until it reached a height of about 10 ft (3 m) above the crown of the pipe. The compaction specification for the structural backfill and this part of the embankment was 100% T-99. A rockfill

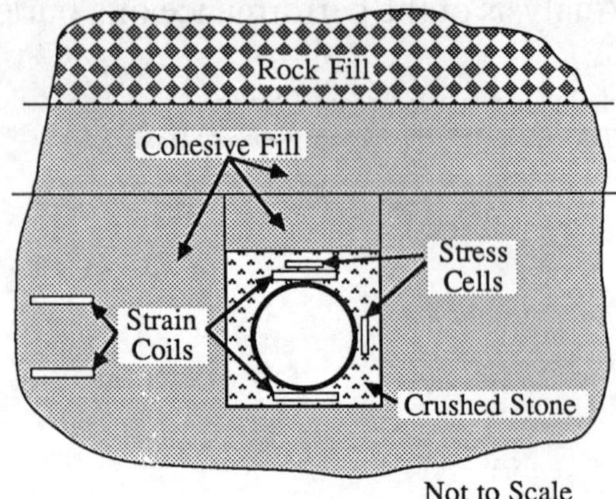

Fig. 3   Location of soil instrumentation

material was then placed in layers of 3 ft (0.9 m) maximum to extend the embankment to a height of 100 ft (30.5 m) above the crown of the pipe.

The instrumentation is described by Adams et al. (1988, 1989) The diameter change was measured using a radial extensometer. The measurements were taken at three locations in each of the instrumented sections and in the central portion of the other sections of the pipe. A video camera provided visual observation of the pipe interior.

The pipe wall circumference change was measured at two locations in each instrumented section using a specially designed gage. The pipe wall temperature and air temperature inside the pipe were measured using thermistors placed in each instrumented section. Bending and compression strains in the pipe wall were measured using a pair of 1-in. (25-mm) diameter inductance coils (Selig, 1975) placed at the crest, the centroid and the trough of the corrugation in a coplanar configuration as shown in Fig. 2. The measurements were taken at the springline level, at the mid-length of each of the instrumented pipe sections.

The instrumentation of the soil around the two pipe sections included soil strain and stress measurements as shown in Fig. 3. The soil normal pressures on the pipe at the springline and at the crown were measured using earth pressure cells located at the mid-length of the instrumented sections. The vertical stress gage is about 2 in. (51 mm) above the pipe crown, and the horizontal stress gage is about 10 in. (254 mm) away from the pipe at springline level. The vertical compression of the structural backfill incorporating the pipe was measured using a pair of 12-in. (305-mm) diameter inductance coils. Additional pairs of coils were used to measure the soil strain in the embankment 16 ft (5 m) away from the axis of the pipe at the same elevation as the pipe.

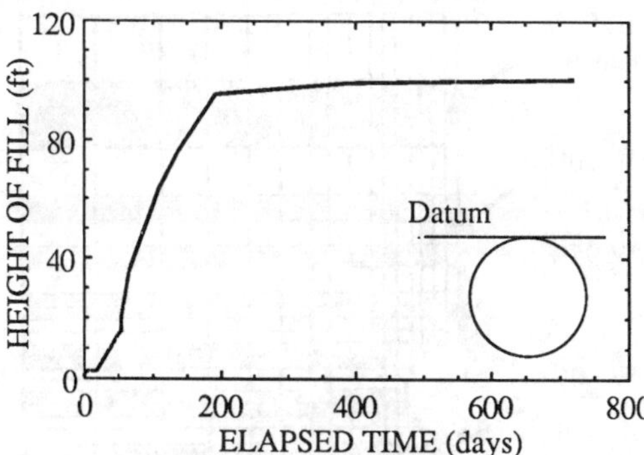

Fig. 4   Fill height change with time

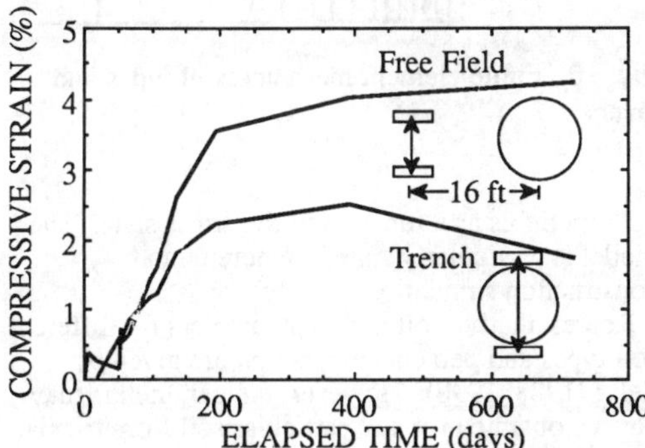

Fig. 5   Increase in vertical soil strain with time

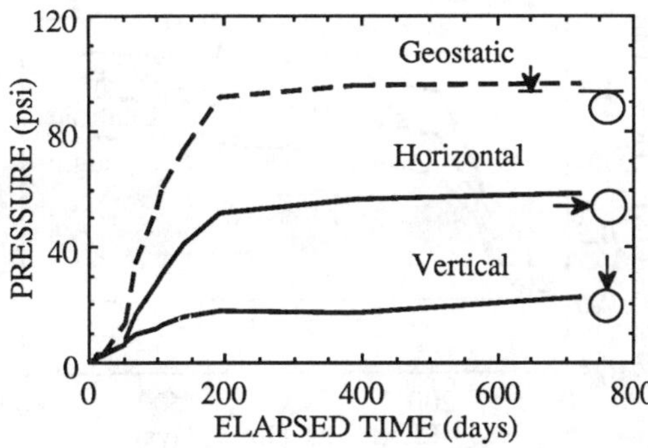

Fig. 6   Increase in earth pressure on pipe with time

3 LONG TERM RESULTS

The results measured during construction are presented and discussed by Adams et al. (1988, 1989). These results will be extended in this paper to include the 2-year period following construction.

The increase in fill height with time is shown in Fig. 4. The embankment reached 95 ft (29 m) above pipe crown 194 days after pipe installation. However, the embankment was not completed until 391 days after pipe installation with fill at 99 ft (30 m) above the pipe crown. The section of the highway above the pipe was paved and completed 722 days after pipe installation. The pipe, at that point, was under 100 ft (30.5 m) of fill measured from pipe crown to the top of the highway pavement.

The pipe temperature at the beginning of construction was about 73°F (23°C). This temperature gradually decreased to about 60°F (16°C) as the embankment was placed.

Fig. 5 shows the trench and free field soil strain variations with time. The average compressive strains 722 days after pipe installation, are 1.7% in the pipe trench and 4.2% 16 ft (5 m) away from the pipe, in the embankment.

Vertical, horizontal and geostatic stress variations with time are shown in Fig. 6. The vertical stress at the crown 722 days after pipe installation is 22 psi (152 kPa), and the horizontal stress at the springline is 58 psi (400 kPa). The vertical free field stress is calculated to be 96 psi (661 kPa). This shows that the vertical stress on the pipe crown has been reduced 77% from the geostatic value. Assuming a lateral at rest earth pressure coefficient of 0.5, the horizontal free field stress is about 48 psi (331 kPa). This free field value is close to the measured value.

Fig. 7 shows the pipe wall strain variation with time at the crest, centroid, and trough of the pipe corrugation at the springline level. After 722 days, the average centroid strain is 1.4% which represents the average wall compression strain. The corresponding average crest and trough compressive strains are 1.7% and 1.1%, respectively. Hence, the bending strain is 0.3%. Fig. 8 shows the shortening of the pipe circumference with time. The circumferential shortening is 1.6% at 722 days after pipe installation.

The average diameter change of the two instrumented sections with time, in four directions, is shown in Fig. 9. After 722 days, the horizontal diameter increased by a value of 0.6%. The vertical diameter decreased by about 4.3%. The visual observation of the pipe interior showed that the pipe was still in good condition.

The long term trends of the various measurements shown in Figs. 5 through 9 approximately follow the trend of the fill height variation with time in Fig. 4.

4 FINITE ELEMENT STUDIES

The field case was analyzed using finite element techniques. Two finite element codes were used, SOILCON and CANDE. The results from each code were obtained and compared to the measured

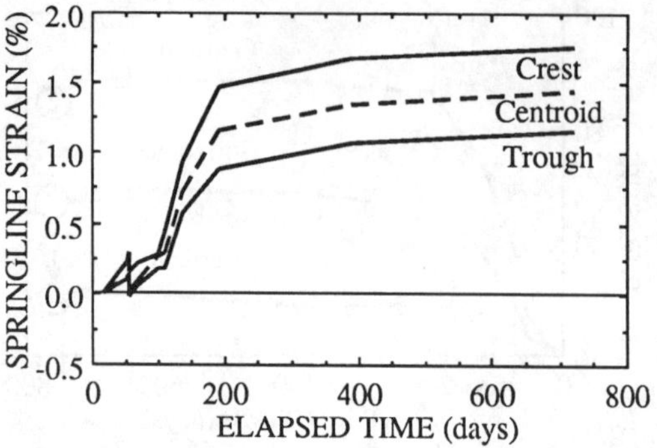

Fig. 7   Compressive wall strain variation with time

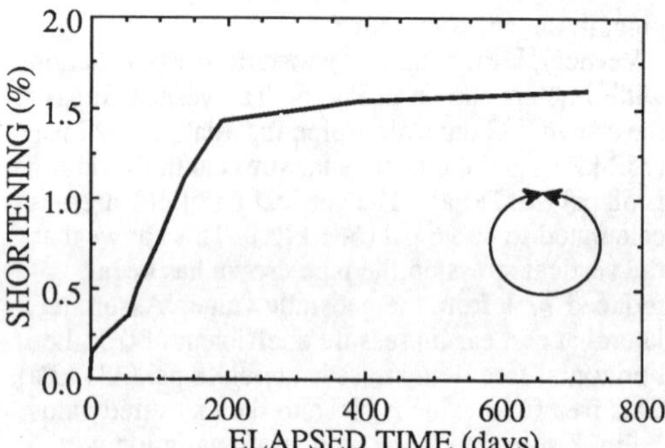

Fig. 8   Shortening of circumference with time

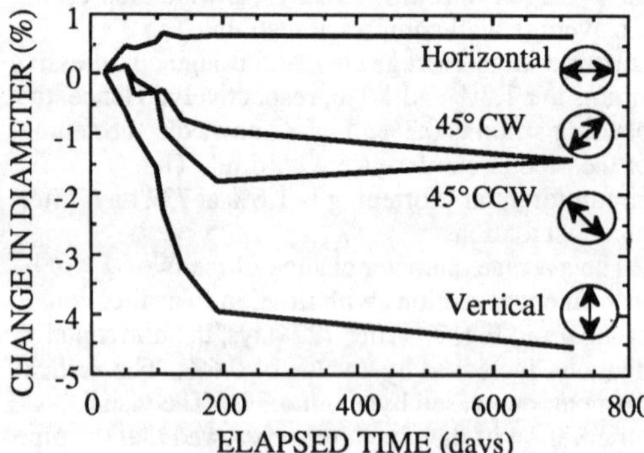

Fig. 9   Pipe diameter change with time

values. The influence of the viscoelasticity of the pipe is illustrated.

4.1 Soil model and parameters

The soil model used for the analysis is described by Selig (1988). It is a hyperbolic soil model representing tangent Young's modulus and tangent

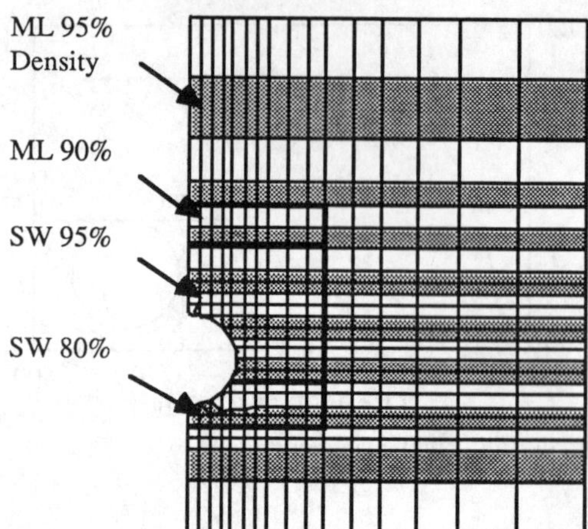

Fig. 10   Finite element mesh and soil types and layers

bulk modulus as a function of the stress state. The model is particularly suited to incremental construction simulation.

Representative soil model parameters for different soil types and percent compaction are given by Selig (1988, 1990). However, the parameters may also be obtained for any particular soil from triaxial tests and hydrostatic compression tests.

The SOILCON code uses this soil model for the analysis of soil-structure interaction problems. This model is also incorporated in CANDE as one of several the user can select to characterize the soil response.

4.2 Description of SOILCON analysis

The program SOILCON (SOIL CONduit analysis) represents flexible buried conduit installation behavior. SOILCON was derived from NLSSIM which was developed by Duncan and his colleagues at the University of California, Berkeley (Kulhawy et al., 1967, Ozawa et al., 1973). Some of the modifications are described by McVay (1982) and Haggag (1989).

The pipe structure is modelled with straight beam-column elements with linear stress-strain behavior. Incremental construction is modelled using load steps to represent placement of a structure or placement of a soil layer. Each increment is iterated twice: first, using values of Young's modulus and bulk modulus at the beginning of the load step, and second, using the moduli values based on the average stresses during the load step.

The finite element mesh used in the analyses is shown in Fig. 10, assuming symmetry about the

Table 1. Hyperbolic soil parameters.

Soil Type	% T-99	γ lb/ft^3 (Mg/m^3)	K	n	R_f	c psi (kPa)		ϕ_o deg.	$\Delta\phi$ deg.	B_i/P_a	ε_u	K_o
SW	95	141 (2.26)	950	0.60	0.70	0	(0)	48	8	187.0	0.014	1.3
	80	119 (1.91)	320	0.35	0.83	0	(0)	36	1	15.3	0.078	0.8
ML	95	127 (2.03)	440	0.40	0.95	4	(28)	34	0	120.8	0.043	1.2
	90	120 (1.92)	200	0.26	0.89	3.5	(24)	32	0	46.0	0.071	0.9

Table 2. Pipe parameters.

Variable	Value	
Young's Modulus, E_p psi (MPa)	110000[a] (758)	22000[b] (152)
Tensile Strength[c] psi (MPa)	3000[a] (21)	900[b] (6)
Moment of Inertia, I_p in.^{4}/in. (mm^4/mm)	0.2 (3277)	
Cross-Sectional Area, A_p in.^{2}/in. (mm^2/mm)	0.353 (9)	
Poisson's Ratio, v_p	0.35	
Average Pipe Diameter, D in. (mm)	25.6 (650)	

[a] Short term values, 0.05 years (AASHTO, 1986b).
[b] Long term values, 50 years (AASHTO, 1986b).
[c] Not used in finite element analysis.

vertical centerline. The soil layers in the lower part of the embankment are placed in 17 construction increments. The pipe is placed after layer 4. The height of fill after all layers are placed is 6.8 ft (2 m) above pipe crown. The remaining 3.2 ft (1 m) of the sand and gravel embankment material, as well as the rockfill material are represented by surcharge pressure applied to the top of the last layer. This pressure is applied in 10 increments corresponding to the increments at which the field data were taken.

The soil types used in the analysis, corresponding to the given installation conditions, are shown in Fig. 10. Four soil zones are specified. The corresponding hyperbolic soil parameters are given in Table 1. The pipe diameter is the average of the inner and the outer diameter. The pipe parameters are listed in Table 2.

4.3 Description of CANDE analysis

CANDE (Culvert ANalysis and DEsign) was first introduced in 1976 for the structural analysis and design of buried culverts (Katona et al., 1976). The code was modified twice (Katona et al., 1981, Musser, 1989). CANDE has a variety of options such as choice of culvert type (corrugated steel, corrugated aluminum, reinforced concrete, and plastic), and choice of analysis or design mode.

For this analysis the level 3 solution of CANDE-89 is chosen (Musser, 1989). A linear stress-strain law for plastic material is assumed. The pipe section properties were modelled using BASIC model where the moment of inertia and the cross sectional area are independent of each other. The PLASTIC pipe model is inappropriate since it calculates the moment of inertia of the pipe wall based on a smooth wall pipe.

The mesh, iterations, soil parameters and pipe parameters were the same as for SOILCON. The only difference between CANDE and SOILCON is in the initial layer placement. For CANDE, the pipe and the soil below it were all placed as the first layer of the construction sequence. The construction was then completed in layers similar to that for SOILCON as shown in Fig. 10. Since all field measurements were initialized after the pipe was placed, the difference in initial layer placement between CANDE and SOILCON did not affect the results.

4.4 Results of analysis

The analysis was done in two stages. The first stage compares the field measurements to the corresponding values obtained from SOILCON and CANDE analyses. The second stage considers time effects, as well as further analysis of the pipe-soil system performance using SOILCON.

The plots of the calculated and measured variation of some of the parameters as a function of fill

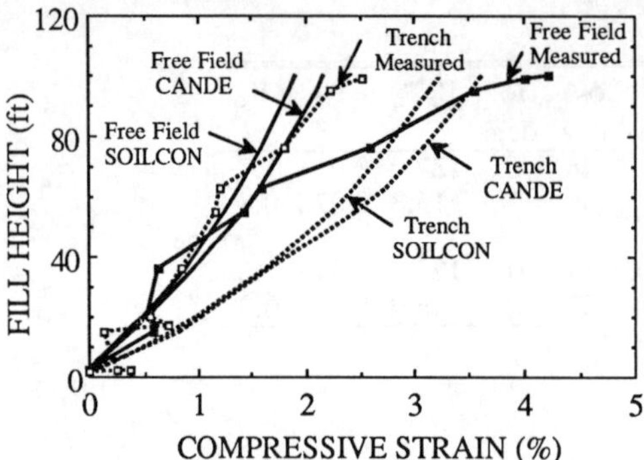

Fig. 11 Variation of measured and predicted soil strain

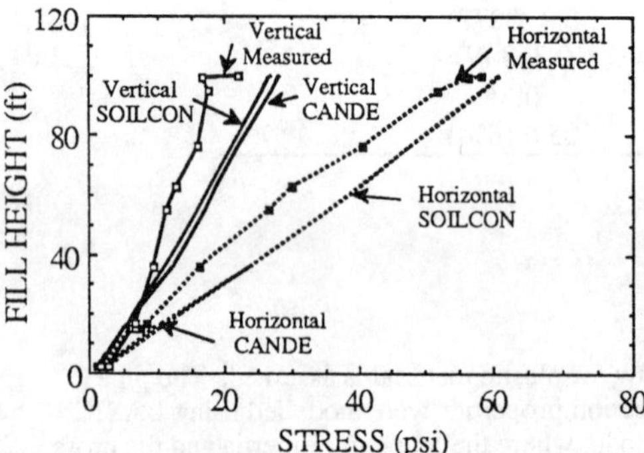

Fig. 12 Variation of measured and predicted soil stress

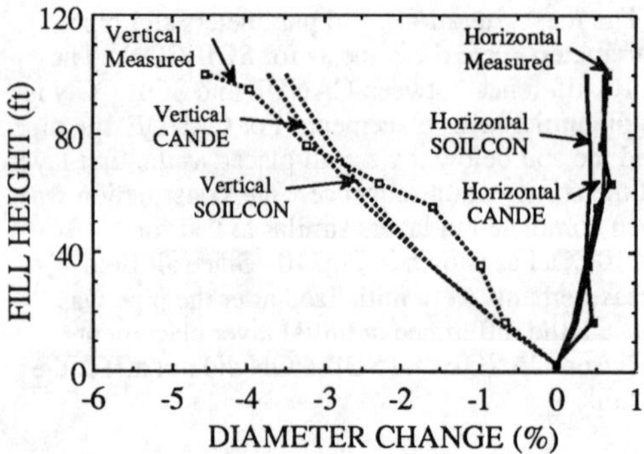

Fig. 13 Variation of measured and predicted pipe diameter

height are shown in Figs. 11 through 13. These variables are referenced to the point where the initial measurements were taken. For pipe diameter change and free field strain measurements the initial value was obtained at 2.5 ft (0.8 m) of cover. For trench strain, the first measurement was taken at 1.5 ft (0.5 m) of cover. The short term pipe modulus

(Table 2) was used for the finite element analysis.

Both codes overestimate the trench strain by 50% after 20 ft (6.1 m) of fill as shown in Fig. 11. Both codes compare well with field measurement for free field strain until about 60 ft (18.3 m) after which the predicted values underestimate the actual values by a maximum of 100%. For both the trench and free field strain, CANDE predicts larger strains than SOILCON does. Soil stresses show a better agreement between the two codes as shown in Fig. 12. However, the predictions overestimate the vertical stresses after 30 ft (9.1 m) by a maximum of 50% and the horizontal stresses after 20 ft (6.1 m) by a maximum of 30%. Fig. 13 shows that the predictions compare well with the measured values of the vertical diameter change. SOILCON underestimates the horizontal diameter change by a maximum of 50% at 100 ft (30.5 m) of fill.

For soil strains, soil stresses and pipe diameter change, CANDE predicts higher values than SOILCON. The predicted wall bending strain, axial strains and circumferential shortening from both codes were the same. Examination of the predicted deformed shape of the pipe at 100 ft (30.5 m) of fill showed that CANDE gives higher pipe settlements. This fact is illustrated in the larger soil strains in Fig. 11.

Due to the viscoelastic nature of the pipe material, time is an important factor in any analysis. Time-dependent representation of the polyethylene stress-strain properties is needed for rigorous modelling. However, the time dependency was approximated in the available static model by reducing the Young's modulus (E_p) of the pipe. AASHTO represents this time effect by giving two values for this modulus, one at 0.05 years and the other at 50 years as shown in Table 2. Since this installation has been in the field for 722 days, the actual value of the modulus is between the two AASHTO values. These proposed AASHTO values as well as values from a laboratory test on the pipe were used to estimate the variation of E_p with time. At 722 days, the modulus of the pipe was estimated by interpolation to be 35000 psi (241 MPa). Finite element calculations were carried out using this value together with the 0.05 and 50 year moduli. The results are compared to those from the field in Figs. 14 through 16.

The figures show that decreasing the modulus by 68% decreases the vertical stress by 17%, reduces the horizontal diameter increase by 170%, and increases the vertical diameter decrease by 29%. The results at 100 ft (30.5 m) of fill for a pipe modulus of 35000 psi (241 MPa) seem to give better agreement with the measured values for the vertical stresses and vertical diameter change. However, this is not the case for the horizontal diameter change.

The distributions of the normal and shear stresses applied by the soil on the pipe at 100 ft (30.5 m) of

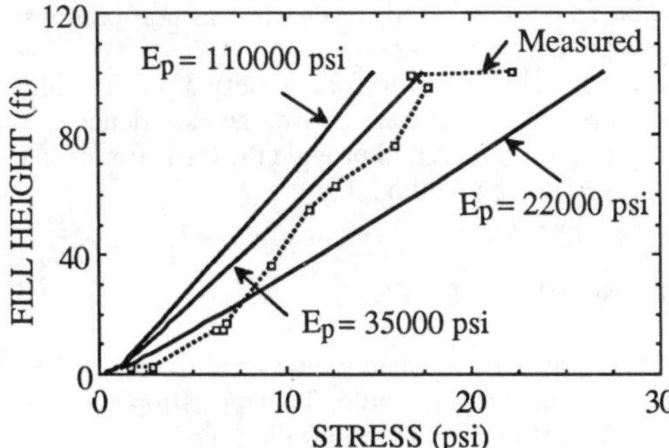

Fig. 14 Effect of E_p on vertical stress

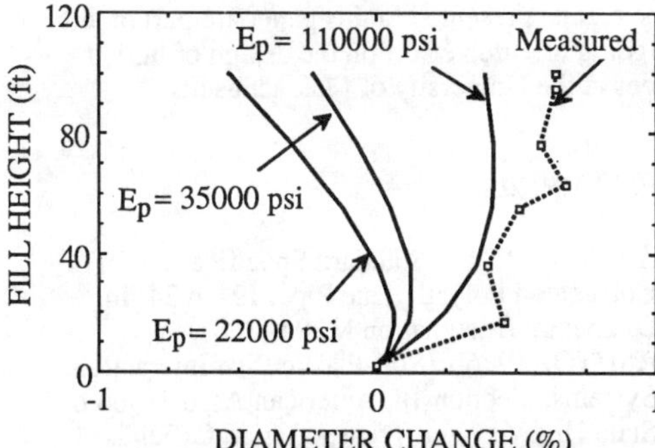

Fig. 15 Effect of E_p on horizontal diameter change

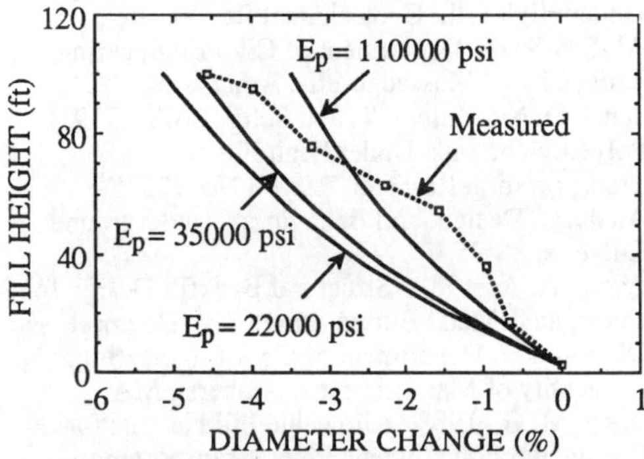

Fig. 16 Effect of E_p on vertical diameter change

cover using a pipe modulus of 35000 psi (241 MPa) are shown in Fig. 17. The measured stresses are higher than those obtained from the analysis. This demonstrates the local arching effect of the structural backfill directly around the pipe.

Arching is defined as the redistribution of free field stresses over a buried structure. It occurs as a result of differences in stiffness properties of the structure and the soil around it. In the case of the buried HDPE pipe, the arching factor determined

from the measured and free field stresses is 0.77. The same factor can be determined using the hoop thrust values. At 100 ft (30.5 m) of fill the thrust at the springline is estimated from the applied pressure and the pipe diameter as 1230 lb/in. (215 kN/m). The corresponding value from the finite element analysis at 722 days is 280 lb/in. (49 kN/m). Thus, the arching factor using these thrust values is also 0.77.

The distributions of the bending and compressive hoop stresses at 100 ft (30.5 m) of fill using a pipe modulus of 35000 psi (241 MPa) are shown in Fig. 18. These stresses are calculated from the moment and thrust, respectively, as obtained from the finite element analysis. Using these stresses, the bending strain at the springline is 0.8% and the compressive hoop strain is 2.7%. The measured values of the bending and hoop strains are 0.3% and 1.4%, respectively.

Fig. 18 shows that the compressive hoop stress at the springline is 945 psi (6.5 MPa) and the maximum value is 1580 psi (10.9 MPa). The allowable long term limit for the hoop stress given by AASHTO is 900 psi (6 MPa) (AASHTO, 1986b). Hence, the stress at the springline is close to the allowable limit whereas the maximum stress greatly exceeds the long term allowable limit.

Design based on CANDE analysis and AASHTO procedures predict that the short term maximum allowable fill height for the pipe under study is 63 ft (19 m) (Katona, 1988). This prediction is conservative since the design procedures are based on tensile properties while gravity flow buried plastic pipes are under compression. Thus, the design stress limit values are probably quite conservative.

5 SUMMARY AND CONCLUSIONS

A field test was conducted to evaluate the performance of a 24-in. (610-mm) diameter HDPE pipe buried under 100 ft (30.5 m) of fill. Finite element analysis was used to predict the pipe performance and compare the predictions with the measured values.

At 100 ft (30.5 m) of cover 722 days from pipe installation, the trench material was found to be less compressible than the free field soil. The trench compression was 1.7% and the free field soil compression was 4.2%. The vertical stress on the pipe crown in the trench was about 23% of the free field value. This indicated an arching factor of 0.77. The horizontal stress of 58 psi (400 kPa) in trench was close to the free field horizontal stress. The pipe wall compressive hoop strain at springline level was 1.4% and the bending strain at springline level was 0.3%. The circumferential strain of 1.6% was

101

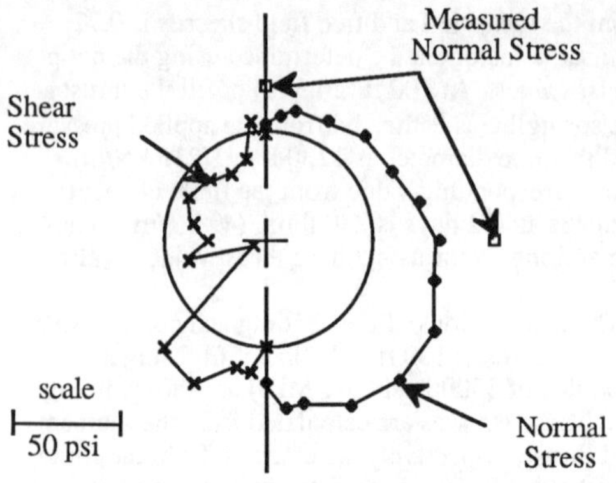

Fig. 17   Distribution of shear and normal stresses

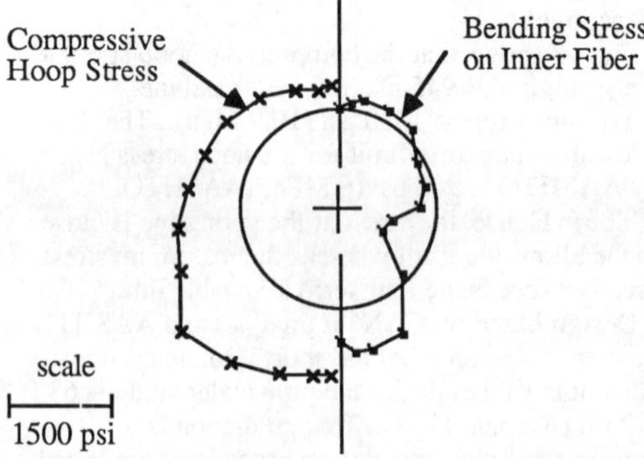

Fig. 18   Distribution of hoop and bending stresses

relatively close to the compressive wall strain. This circumferential strain produced a vertical diameter decrease of 4.3% and a horizontal increase by 0.6%. The visual observation of the pipe interior showed no pipe distress and a round appearing pipe shape. The long term measurements were found to be following the same trend as the fill height change with time.

The short term analysis showed reasonable agreement with the measured values during construction. SOILCON and CANDE gave similar predictions. The major difference between the two codes is that CANDE predicts greater soil compression. The long term analysis compared well with the long term results. The analysis was useful in predicting pipe wall stresses. However, a time-dependent model should be considered to properly model the long term response of the soil-pipe system.

The predicted maximum compressive hoop stress of 1580 psi (10.9 MPa) is higher than the long term allowable limit, even though the pipe is performing satisfactorily. This is because the design procedures are based on tensile properties whereas non-pressure pipes are under compression.

Further field studies under a variety of installation conditions are desired to confirm the experience shown in this paper, and provide the basis for an improved design methodology.

ACKNOWLEDGEMENTS

The long term data collection was sponsored by the Pennsylvania Department of Transportation and the U. S. Department of Transportation, Federal Highway Administration. The pipe installation and data collection up to 95 ft (29 m) of fill were sponsored by Advanced Drainage Systems, Inc. The results presented in this paper are part of an ongoing research effort on the design of buried pipes at the University of Massachusetts.

REFERENCES

AASHTO. 1986a. Standard Specification for Corrugated Polyethylene Pipe, 12- to 24- in. Diameter. Designation M294-86.

AASHTO. 1986b. Soil-Plastic Pipe Interaction Systems. Section 18, American Association of State Highway and Transportation Officials, Flexible Culvert Committee, San Francisco, CA.

Adams, D. N., Muindi, T. and Selig, E. T. 1988. Performance of High Density Polyethylene Pipe Under High Fill. Geotechnical Report No. ADS88-351F, Department of Civil Engineering, University of Massachusetts, Amherst.

Adams, D. N., Muindi, T. and Selig, E. T. 1989. Polyethylene Pipe Under High Fill. Transportation Research Record No. 1231, Analysis, Design, and Behavior of Underground Culverts: 88-95.

Haggag, A. A. 1989. Structural Backfill Design for Corrugated-Metal Buried Structures. Doctoral Dissertation, Department of Civil Engineering, University of Massachusetts, Amherst, MA.

Katona, M. G. 1988. Allowable Fill Heights for Corrugated Polyethylene Pipe. Transportation Research Record No. 1191, Culverts and Tiebacks: 30-38.

Katona, M. G., Smith, J. M., Odello, R. S. 1976. CANDE - A Modern Approach for the Structural Design and Analysis of Buried Culverts. Report No. FHWA-RD-77-5, Federal Highway Administration, Naval Civil Engineering Laboratory, Port Hueneme, CA.

Katona, M. G., Vittes, P. D., Lee, C. H., and Ho, H. T. 1981. CANDE-1980: Box Culverts and Soil Models. Report No. FHWA/RD-80/172, Federal Highway Administration, Washington D. C.

Kulhawy, F. H., Duncan, J. M., and Seed. 1967.
Finite Element Analysis of Stresses and
Movements in Embankments During
Construction. Report No. TE 69-4, Office of
Research Services, University of California,
Berkeley, CA.

McVay, M. C. 1982. Evaluation of Numerical
Modelling of Buried Conduits. Doctoral
Dissertation, Department of Civil Engineering,
University of Massachusetts, Amherst, MA.

Musser, S. C. 1989. CANDE-89 User Manual.
Report No. FHWA-RD-89-169, Federal Highway
Administration, Mclean, VA.

Ozawa, Y. and Duncan, J. M. 1973. ISBILD: A
Computer Program for Analysis of Static Stresses
and Movements in Embankments. Report No.
TE-73-7, Office of Research Services, University
of California, Berkeley, CA.

Selig, E. T. 1975. Soil Strain Measurement Using
Inductance Coil Method. Performance
Monitoring for Geotechnical Construction, ASTM
STP 584: 141-158.

Selig, E. T. 1988. Soil Parameters for Design of
Buried Pipelines. Proceedings of the Conference
on Pipeline Infrastructure, Boston, MA, American
Society for Civil Engineers: 99-116.

Selig, E. T. 1990. Soil Properties for Plastic Pipe
Installations. Buried Plastic Pipe Technology,
ASTM STP1093, G. S. Buczala and M. J.
Cassady, Editors, American Society for Testing
and Materials, Philadelphia.

Structural Performance of Flexible Pipes, Sargand, Mitchell & Hurd (eds) © 1990 Balkema, Rotterdam. ISBN 90 6191 165 6

Structural performance of three foot corrugated polyethylene pipe buried under high soil cover

Reynold K.Watkins
Utah State University, Logan, Utah, USA

ABSTRACT: Corrugated, high density polyethylene pipes of 24 inch diameter and less, perform successfully as buried culverts. Because 36 inch pipes are now available, a test was conducted to discover if 36 inch (914 mm) pipes, scaled up from the smaller diameters, would perform as well structurally under high soil cover. The result was positive. The soil was compacted fine sand. A recommended conservative performance limit for the pipe is dimpling at 9:00 and 3:00 o'clock at ring deflection of about 5%.

1 INTRODUCTION

Corrugated high density polyethylene pipes have proved successful as buried drain pipes and culverts. To the present time, diameters have been 24 inches (610 mm) and less. New technologies for manufacture make possible the production of larger diameters. Questions arise as to the structural performance of larger diameters, especially when buried under adverse loading conditions such as high soil cover or heavy surface loads over minimum soil cover. A test was performed in the large USU soil cell to investigate the structural performance of 36 inch (914 mm) diameter corrugated high density polyethylene pipe under high soil cover. Because smaller diameter pipes perform successfully in marginal pipe-zone backfill, the soil selected for the 36 inch pipe test was also marginal -- near the lower limit allowed by the California Department of Transportation, Standard Specifications number 19-3.0 on structure backfill. The pipe-zone backfill was fine sand with about 20% silt. It is blow sand -- typical of much of the surface soil in the United States.

In practice, pipes are usually buried in trenches. Occasionally they are laid on plane surfaces and soil is placed over them creating an embankment. If the soil is not compacted, or is only marginally compacted, the embankment soil load is more adverse than the trench load. For this reason, in order to investigate worst conditions, soil was placed in an embankment over the test pipe.

Figures 1 and 2 show the large USU soil cell. It is 15 ft (4.6 m) wide 22 ft (6.7 m) long, and 18 ft (5.5 m) high. It is basically a steel plate cylinder with an elliptical cross section, truncated on top such that ten load beams can be pinned

into place for applying load. Load is applied by hydraulic jacks --five on each load beam. A section of test pipe is buried in the soil cell. When the soil cover is equal to one pipe diameter or more, the vertical soil pressure applied by the 50 hydraulic jacks simulates high soil cover. An access pipe allows personnel to enter the pipe during loading in order to observe and monitor structural performance of the pipe.

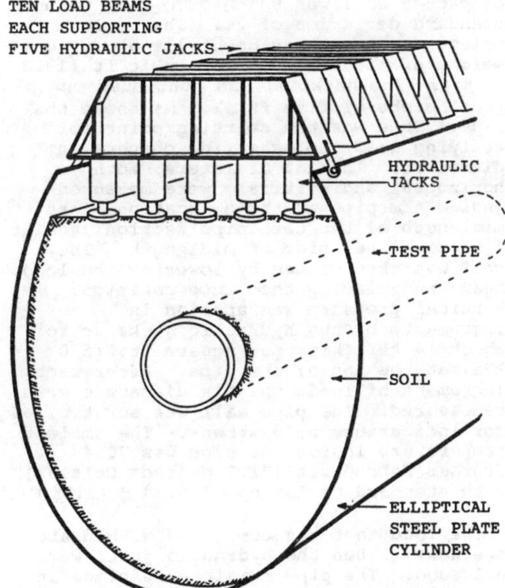

Figure 1. Drawing of the cross section of the large USU soil cell showing the elliptical steel cylinder with load beams on top for applying vertical pressure.

105

Figure 2. Large USU soil cell ready for pressurizing the hydraulic jacks and showing the access pipe in the foreground through which test pipe performance is monitored.

2 PROCEDURE

A test section of corrugated polyethylene pipe, 36 inch (914 mm) inside diameter and twenty ft (6.1 m) long was placed in the cell on a level compacted soil plane. This is a worst case for bedding -- yet it is not uncommon in practice. The pipe-zone backfill was placed in 8 inch lifts and compacted by means of hand operated rammers. The average density of the compacted soil was 93% AASHTO T-99 with a standard deviation of 2%. The optimum moisture content was about 11% at dry unit weight of about 126 lb per cubic ft (19.8 kN/m^3). The backfill was continued on up to a height of five ft (1.5 m) above the pipe. This was the starting point for applying load and measuring changes in diameters. Initial diameters, both horizontal and vertical, were measured inside the pipe at three locations; at midlength of the test pipe section, and at five ft either side of midlength. The cell was then closed by lowering the load beams and pinning them into position. Vertical pressure was applied in increments by the hydraulic jacks up to about 14 ksf (kips per square ft) (670 kPa) at the top of the pipe. After each increment of load, the six diameters were remeasured. The pipe wall was scrutinized for indications of distress. The ambient temperature inside the pipe was 70.4 degrees Fahrenheit (21.3 degrees Celsius) with standard deviation of 6.3° F (3.5° C).

The load test was completed within about one hour. Then the hydraulic jacks were unloaded. The pipe remained in place in the cell for two days after which it was reloaded and unloaded twice to 14 ksf (670 kPa) in order to observe the effect of stress relaxation on the soil-structure interaction.

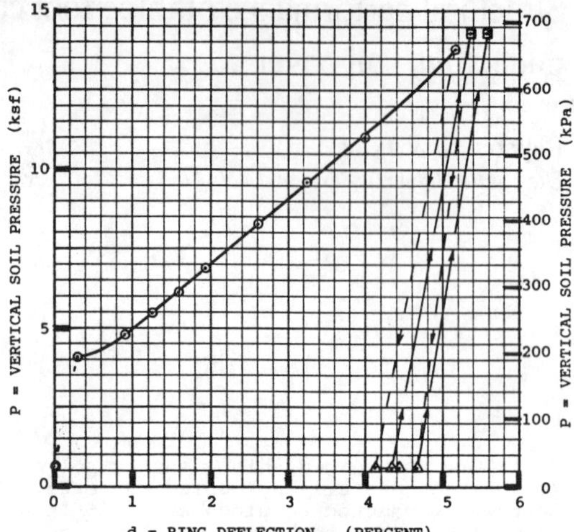

Figure 3. Relationship between vertical soil pressure and ring deflection showing two additional cycles, two days later, of reloading from 0.58 kips per square ft (27.8 kPa) to 14.34 ksf (686.6 kPa). Ring deflection is the decrease in vertical diameter divided by the diameter 37.61 inches (955 mm) of the neutral surface of the pipe wall. Ring deflection starts from zero at 0.58 ksf.

3 RESULTS

The results of the test are shown on Figure 3 which is a plot of vertical soil pressure P at the top of the pipe, as a function of vertical ring deflection d at midlength of the pipe. Ring deflection was the change in vertical diameter divided by the original pipe diameter.

Because of the compaction of sidefill soil, there was no significant ring deflection at less than about 4 ksf (192 kPa). The first indication of distress in the pipe wall occurred at a vertical soil pressure of 13.0 to 13.5 ksf (622 to 646 kPa). At 9:00 and 3:00 o'clock, the cylindrical surface inside the pipe showed a slight deformation that could be a precursor to the formation of plastic hinges. The observation was facilitated by moving a light source until shadows revealed very shallow dimpling of the pipe wall. This indication of distress occurred at ring deflection of about d = 5%.

where
d = Δ/D = vertical ring deflection
Δ = decrease in vertical diameter
D = diameter of the neutral surface of the pipe wall

The rate of loading was so fast that the plastic pipe had little time to relax. Some relaxation is evident in the two

reload cycles performed two days after the
initial test. Had the loading been
slowed, say to a period of a week rather
than an hour, the dimpling could only have
appeared at some vertical soil pressure
greater than 13.5 ksf (646 kPa). Of
course, field construction of high
embankments proceeds at a slow rate.
Stresses in the pipe have time to relax.

Dimpling may be identified by
conservative design engineers as a
reasonable performance limit, and so, the
basis for design. However, it is not a
precursor for collapse. Because the
stresses in the pipe wall relax under
constant ring deflection, collapse is not
incipient even though the pipe wall is
dimpled. Consequently, design of the pipe
should be based on 5% ring deflection
allowable. Because stresses in the pipe
relax, design should be based on quick
modulus of elasticity and early strength
-- not 50 or 100 year strength under
persistent stresses. It is of interest
that when the pipe was removed from the
soil cell, it rebounded and the dimpling
disappeared by the following day.

It is essential that the pipe-zone
backfill be good granular soil with a
large soil friction angle. Viscous soil
(mud), causes persistent external
hydrostatic pressures and, therefore,
persistent stresses for which the long-
term strength should be used for design.
Moreover, under the worst circumstances,
external hydrostatic pressure could cause
pipe collapse. Of course, this applies to
all flexible pipes -- not just corrugated
polyethylene pipes.

The possibility of collapse is greatly
reduced if the shape of the pipe ring is
held close to circular during placement of
the pipe-zone backfill. This can be
assured by care in placing soil around the
pipe; and by compacting the side fill soil
enough that vertical compression of the
sidefill soil will not exceed the
allowable ring deflection of the pipe.

One of the attractive characteristics of
flexible pipes is their ability to conform
with the soil and so relieve the pipe of
soil load concentrations. The phenomenon
tends toward a soil tunnel in which the
pipe is a tunnel liner. Soil load is
further relieved by relaxation of stresses
in the pipe wall.

There is no reason why corrugated
polyethylene pipes of three ft diameter
cannot perform structurally under high
soil cover provided that a good granular
pipe zone backfill is carefully placed and
compacted.

Structural Performance of Flexible Pipes, Sargand, Mitchell & Hurd (eds) © 1990 Balkema, Rotterdam. ISBN 90 6191 165 6

The design and construction of culverts using Controlled Low Strength Material – Controlled Density Fill (CLSM-CDF) backfill

William E. Brewer
Ohio Northern University, Ada, Ohio, USA

ABSTRACT: The history and development of Controlled Low Strength Material (CLSM) is referenced. Various applications for CLSM, other than flexible pipe backfill are noted. Future research on CLSM mixtures is suggested.

The design and construction of flexible pipes are influenced by the quality of the backfill that can be achieved around the pipe. The use of a (CLSM) assures proper pipe bedding and results in improved structural capabilities and cost of a flexible pipe installation. Four general areas: design, materials, construction, and economics for the use of CLSM as a flexible pipe backfill are reviewed.

Values for the modulus of soil reaction E' are presented for CLSM and their effect on horizontal deflections compared to conventional backfill. For the same E' value, CLSM and conventional backfills, CLSM backfill could result in a 50% reduction in horizontal deflection because of bedding and deflection lag factors. Values for E' can be determined for various CLSM mixtures. Trial mixture design information is presented with field control procedures. Economics of using CLSM backfill as opposed to conventional backfill is examined and a case history is presented.

1 HISTORY

CLSM is the acronym for Controlled Low Strength Material. This acronym was developed by an American Concrete Institute committee (ACI 229) for its development work with low strength materials. This committee has defined, for its use, low strength material to be material that consists of: portland cement, water, and selected filler material such as fly ash and aggregate and possesses a compressive strength of less than 1200 psi. at 28 days. The compressive strength value is just a general rule of the committee and has no engineering significance. CLSM mixtures should not be confused with "lean mix concrete". Lean mix concrete is a term used for reducing the cement factor in a portland cement concrete mixture and is usually designed using portland cement concrete principles. CLSM mixtures are <u>not designed</u> by portland cement concrete principles. CLSM makes use of three engineering areas of study: concrete, ceramics, and soils. Thus one finds references to concrete tests and nomenclature for material being used in soils.

Prior to the organization of the ACI 229 committee (1982) work with low strength materials was being carried on, and still is, by a number of agencies and individuals. It was Joseph Aspdin's work in the 1850's that began the study of portland cement and uses for concrete. He believe he had duplicated a stone similar to that mined on the Isle of Portland.

Around 1900 scientists and engineers were developing special areas of material study: soils, steel, cement, concrete, and wood. This specialization hindered the

development of materials such as
CLSM. Contractors and ready mixed
concrete producers adjusted their
concrete mixtures for what was
considered "low strength" uses: mud
mats, general structural fills, and
pavement bases. They did not
concieve any possible extended uses
for low strength materials.

In the early 1970's some
individuals considered the
possibility of uses for low strength
materials (0 to 1200 psi.). These
considerations led to the formation
of a company known as K-Krete, Inc.
and the marketing of K-Krete
mixtures through franchised agents.
Specially designed K-Krete mixtures
were designed for: trench backfill
(CDF), pavement base (CPB),
structural fill (CSF), thermal fill
(CTF) and so on. CDF stood for
controlled density fill, CPB for
controlled pavement base, CSF
referenced controlled structural
fill, and CTF designated controlled
thermal fill. The control signified
that the mixture was designed and
had manufacturing and construction
control requirements.

By 1979 the concrete/construction
market had many products being
marketed as a K-Krete mixture or
equal. Each producer had their own
brand name and seemingly very little
information as to proper use or
control. It was evident that
technical information, developed by
many individuals, was not being
disseminated. Therefore, the
American Concrete Institute (ACI)
formed an international committee
dealing with the subject of CLSM in
1983.

2 FLEXIBLE PIPE DESIGN TECHNIQUES USING CLSM-CDF

2.1 Previous Research

Previous research publications and
textbooks (1, 2, 3, 4, 5, 6) contain
a considerable amount of design
information for flexible pipes.
They all stress a common concern,
adequate engineered backfill around
the flexible conduit. The American
Iron and Steel Institute states (6):

"Stability in a soil-structure
interaction system requires not only
adequate design of the structure
barrel, it also presumes a well
engineered backfill. Performance of
the flexible conduit in retaining
its shape and structural integrity
depends greatly on selection,
placement and compaction of the
envelope of earth surrounding the
structure and distributing its
pressures to the abutting soil
masses."

2.2 Iowa Formula and Modulus of Soil Reaction (E')

The Hartley and Duncan (5) paper
provides an excellent review of the
development of the Iowa formula and
the Modulus of Soil Reaction (E').
This paper also references work by
Howard (1) and his work in field
measurement for deflection and the
determination of E' for various soil
conditions.

Numerous papers have been written
regarding design procedures for
flexible conduits. The majority of
these papers, and this paper,
reference the use of the Iowa
formula. The Iowa formula considers
deflection a critical design factor
and designers recommend setting
deflection limits to a % of the
nominal diameter. This paper uses
the Iowa formula for making
comparisons between conventional
backfill and CLSM-CDF backfill. The
Iowa formula is expressed as:

$$\Delta X = D_1 \frac{KWr^3}{EI + 0.061\ E'r^3}$$

where:

ΔX = horizontal deflection of the
pipe, in inches.

D_1 = deflection lag factor to
compensate for the volume
change of the soil with time,
no dimension.

K = bedding constant which varies
with angle of bedding,
no dimension.

W = vertical load on the pipe per
unit length, lb./linear inch.

r = pipe radius, in inches.

E = modulus of elasticity of the
pipe material, lbs./sq.in.
I = moment of inertia per unit
length of cross section
of pipe wall, inches^4/in.
E' = modulus of soil reaction,
lbs/sq.in.

The Iowa formula can be rearranged
into the form (5):

$$\triangle X = \frac{D_1 KW}{\dfrac{EI}{r^3} + 0.061 \, E'}$$

This equation can be expressed in
general terms: the horizontal
deflection is equal to the load
factor (D KW) divided by the sum of
the Ring Stiffness Factor (EI/r^3)
and the Soil Stiffness Factor (E').

2.3 Design Comparisons: Conventional Backfill And CLSM-CDF

Various values in the Iowa formula
are examined for design comparsions
between conventional backfill and
CLSM-CDF. Deflection comparisons
for $\triangle X$ will be made to demonstrate
design advantages for using CLSM-CDF
in place of a conventional backfill.

D_1 - the deflection lag factor is
applied to compensate for the volume
change in the backfill with time.
This is an empirical value with a
range of 1.0 to 6.0. Sprangler (1)
recommends a value of 1.5
for soil backfill. Since CLSM-CDF
has not exhibited any volume change
with time, a value of 1.0 may be
used.

K - the bedding constant which
varies with angle of the bedding
around the pipe. Values for K are
0.110 and 0.083 for 0 degree and 90
degree bedding angles respectively
(2). See Figure 1. For comparsion
calculations, the conventional
backfill will have a K value of
0.110 and the CLSM-CDF a value of
0.083 since the CLSM-CDF provides
positive bedding in excess of the 90
degrees. K bedding values will vary
for different angles of α. The use
of CLSM-CDF results in a bedding
angle of 90 degrees in all cases.
The same bedding angle can be
achieved with conventional backfill,
but it would require more labor than
that required for CLSM-CDF.

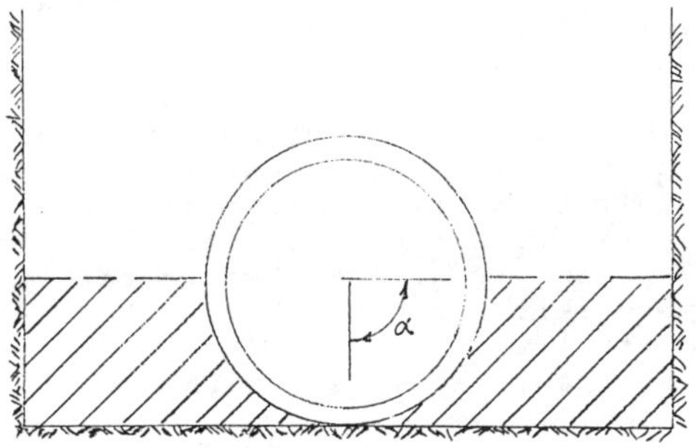

Bedding Angle, α (deg)	Bedding Constant, K
0	0.110
15	0.108
22½	0.105
30	0.102
45	0.096
60	0.090
90	0.083

Figure 1.

(From Soil Engineering (Third Edition)
Spangler and Hardy, Refenerce 2)

W - the load on the pipe, per unit
length, will be considered the same
for both backfill materials.

The Ring Stiffness Factor (EI/r^3)
will be considered the same for both
backfill materials.

The Soil Stiffness Factor (0.061 E')
will be considered the same for both
backfill materials. E' values for
different CLSM-CDF mixtures and soil
characteristics (1) will be
presented later in this paper.

If the above referenced values
were used in the deflection
(horizontal) equation and the
results compared between the
conventional and CLSM-CDF backfills,
the CLSM-CDF backfill would reduce
the deflection by 1/2 for the same
E' values for the bedding conditions
referenced.

3 CLSM MATERIAL REQUIREMENTS: DESIGN, TRANSPORTING TO SITE, PLACEMENT, FIELD TESTING, AND QUALITY CONTROL

3.1 Materials For CLSM-CDF

CLSM-CDF is a controlled density
fill made by the cementation of

materials: portland
cement, water, and selected filler
material such as fly ash and
aggregate. The proper combining of
these materials will result in a
product that will achieve a
compressive strength of less than
100 psi at 28 days and would permit
removability at some later date. If
removability is not to be
considered, higher strengths could
be used.

The CLSM-CDF material should have
good flowability to allow for
uniform bedding amd placement around
the conduit. Good flowability also
means that the material will seek
its own level, during placement,
while all mixture components remain
in suspension.

3.2 Standard Laboratory Reference
   Mixture

Trial mixtures for CLSM-CDF should
be made when considering new
aggregate filler sources. As noted
previously, for a CLSM-CDF the 28
day compressive strength should be
less than 100 psi if removability is
to be considered. A starting point
for trial mixtures, to achieve this
strength, is:

Portland Cement (Type I)	100 lb.
Fly Ash (F)	300
Aggregate Filler	
(Fine Aggr.)	2600
Water	600

This standard trial mixture has
been used in ODOT research for pipe
arch and long span arch projects.
Aggregate filler samples have been
collected from all parts of the
United States and some of the data
have been used in this report for
the determination of the modulus of
soil reaction (E').

3.3 Design Of CLSM-CDF For E' Values

Hartley and Duncan (5) suggest that
the following equation provides good
correlation between E' and the
constrained soil modulus Ms.

$$E' = Ms = \frac{Es\ (1-Vs)}{(1+Vs)\ (1-Vs)}$$

Table 1

Where:

S = stress (lb/sqin)   e = strain (in)

Es = Young's modulus Vs = Poisson's ratio

E' = modulus of soil reaction (lb/sqin)

MODULUS OF SOIL REACTION (E') - CONTROLLED LOW STRENGTH MATERIA

Sample #	s	e	Es	E' for values of Vs		
				0.2	0.3	0.4
28	4	0.075	53	59	72	114
49	20	0.035	571	635	769	1224
53	21	0.065	323	359	435	692
50	25	0.053	476	529	641	1020
41	45	0.055	818	909	1101	1753
37	48	0.095	505	561	680	1083
39	49	0.055	891	990	1199	1909
31	63	0.120	525	583	707	1125
38	64	0.125	512	569	689	1097
29	73	0.095	768	854	1034	1647
32	74	0.135	548	609	738	1175
27	81	0.120	675	750	909	1446
34	85	0.110	773	859	1040	1656
40	95	0.090	1056	1173	1421	2262
33	129	0.085	1518	1686	2043	3252
35	139	0.113	1236	1373	1663	2648

where:

    Es = Young's modulus
    Vs = Poisson's ratio

Laboratory data collected as part
of an ODOT sponsored study for
CLSM-CDF are shown in Table 1. This
table contains: the sample number,
stress (s), strain (e), Young's
modulus (Es), and E' for three
values of Poisson's ratio. Table 1
has been prepared for various types
of aggregate filler. There is
correlation between the CLSM
compressive strength, Es, and E'
values. See Figure 2 for the
relationship between Young's modulus
(Es) and CLSM compressive strengths.

The normal range for Poisson's
ratio for engineering materials is 0
to 0.5 (2). Table 2 shows the
Poisson's ratio for various soil
types (8). Poisson's ratio for
concrete is in the range of 0.11 to
0.21 (7). The estimated range for
the CLSM-CDF would be 0.20 to 0.40
depending on the aggregate filler
being used. Three Poisson ratio
values: 0.2, 0.3, and 0.4 are
included in Table 1. The

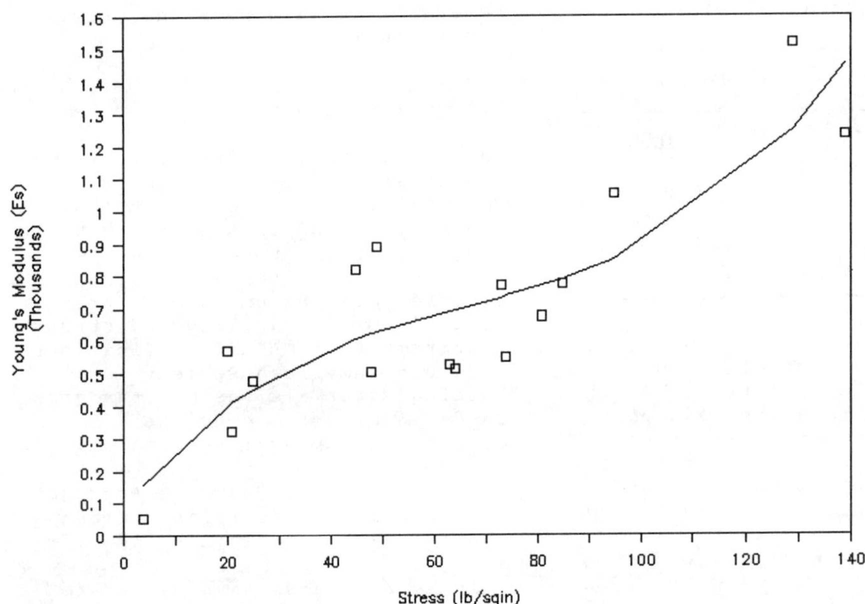

Figure 2

relationship between Poisson ratios, Es, and E' are shown in Figure 3. The higher the Poisson ratio the larger the E' value.

E' values in Table 1, when compared to E' values in Table 3 (1), demonstrate that CLSM-CDF can be manufactured to meet and surpass any soil-type pipe bedding desired. While the Hartley and Duncan (5) paper notes that E' increases with depth of cover for conventional backfills as shown in Table 4, it is believed that this is not the case with CLSM-CDF backfills. This is because the CLSM-CDF does not consolidate, with time, after its initial subsidence. Future studies will investigate this belief.

3.4 Transporting To Construction
    Site

In the majority of the projects, CLSM-CDF is transported to the construction site in ready mixed concrete trucks. The mixer on the ready mixed concrete truck keeps the CLSM-CDF mixture from segregating during

transporting. Studies are currently being proposed to evaluate other transporting and mixing procedures for CLSM mixtures, specifically site mixing for large volume backfill requirements.

3.5 Placement Of CLSM-CDF

The CLSM-CDF material may be poured directly from the truck. No vibration or compaction effort is required. The material should be poured on both sides of the structure at the same time and rate to maintain equal horizontal forces. Flotation of the culvert is possible if the CLSM-CDF is poured too rapidly. Flotation calculations should be made to determine if vertical restraint is required for the planned rate of CLSM-CDF placement. Headwalls or end dams are required to restrain the material at each end of the culvert.

Since CLSM-CDF does not support vegetation a soil cover is required over the backfill material on slopes. Roadway surface material can be placed directly on the CLSM-CDF.

Table 2
Table from "Soil Technology And
Engineering Properties Of Soils"
Reference 8.

Soil Type	Poisson's Ratio
Clay, saturated	0.50
Clay with sand and silt	0.30-0.42
Clay, unsaturated	0.35-0.40
Loess	0.44
Sandy soil	0.15-0.25
Sand	0.30-0.35

Another advantage of using
CLSM-CDF is that the trench width
for the culvert can be reduced.
Conventional backfill requires
enough width outside the culvert to
achieve adequate bedding. The
CLSM-CDF mixture will flow around
the bottom of the culvert
with a minimum, work area, trench
width.

3.6 Field Testing

CLSM-CDF can be tested for
compressive strength and
flowability. Testing for
compressive strength is performed as
set forth in ASTM C39 with
modifications for cylinder size and
rodding requirements. The cylinder
size can be either 4x8 or 6x12
inches (diameter x length). Samples
should not be rodded. Since the
compressive strength is low, the
cylinders are not stripped until
ready for testing.

The flow test consists of filling
a 3x6 inch open-end cylinder mold to
the top with the CLSM-CDF mixture
and striking off the top of the
cylinder so that the mixture is
level. This striking off may not be
necessary for good fluid mixtures.
Any excess should be cleaned off
from around the base of the cylinder
and the cylinder pulled straight up.
The approximate diameter of the
mixture's spread is measured. Good
flowability is achieved when the
diameter is approximately 8 inches.

Some agencies request unit weight
tests. This would be important if a
specific density was the major

design requirement. For general
backfill, compressive strength and
removability have been the prime
considerations.

3.7 Quality Control

Quality control of CLSM-CDF mixtures
requires: uniformity of the
materials, proper equipment charging
and mixing, and transporting to the
project site. It cannot be over
stressed that materials for CLSM-CDF
mixtures must be uniform to provide
control for flowability and strength
requirements. CLSM specifications
have used American Society for
Testing Materials (ASTM) standards
as designated for concrete
components for quality control:

    ASTM C33   - Concrete Aggregate
    ASTM C150   - Portland Cement
    ASTM C618   - Fly Ash

These standards have helped to
provide for uniformity in CLSM
mixtures. Materials not meeting
these standards have also been
successfully used. Use of non-ASTM
materials will require laboratory
testing to determine flowability,
removability, and strength of the
mixture. The field testing
requirements for CLSM-CDF acceptance
has been minimal. The reason for
this is that the engineering
properties can be determined prior
to use.

4   ECONOMIC CONSIDERATIONS FOR
    USING CLSM-CDF IN PLACE OF
    CONVENTIONAL BACKFILL

It is difficult to review the
economics of construction costs
because of variations in materials,
contractors, inspection, and
technical knowledge from one area to
another. Even with this in mind,
cost studies must be made when
considering the use of new
materials. The use of CLSM-CDF, in
all applications, requires a cost
feasibility study. If a CLSM
mixture isn't cost effective it
shouldn't be used. To better
understand the economics of using
CLSM-CDF, as a backfill in place of
conventional backfill, a case
history for a culvert will be

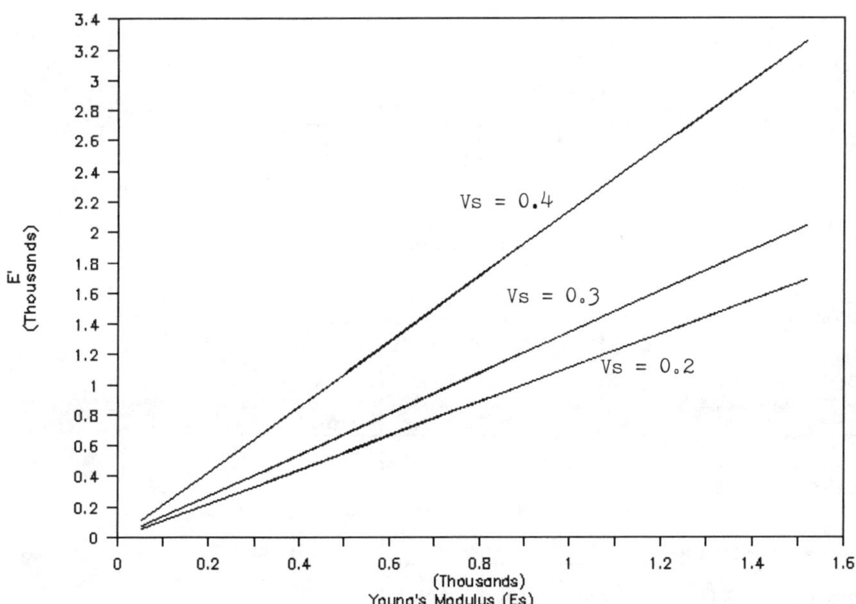

Figure 3

presented. The presentation will include cost information related to materials and construction methods. Naturally, changes in unit costs would result in a different selection for the culvert's backfill.

This case history is based on average estimated costs. As noted previously, if there were changes in material costs or production rates the results could change in favor of granular backfill. This case history also outlines the major components for a cost feasibility study. Other factors not included in this review would be: time required for reopening roadway to traffic, safety related to not having workers in a trench for backfilling, and the assurance that the trench is uniformily backfilled.

The CLSM-CDF backfill will be supplied from a ready mixed concrete producer. The unit cost of the CLSM-CDF has been adjusted for yield and mixture changes for conventional concrete mixtures. The same overhead costs, for concrete, have been used for the CLSM-CDF backfill. Ready mixed concrete producers have not taken the initiative, in general, in seeking cost study information for the use of CLSM-CDF. Cost variances of 200% have been recorded in the United States for similar CLSM-CDF mixtures. A "rule of thumb" has been that CLSM-CDF would cost and half the cost of a 6 sack concrete mixture.

5 Case History - Culvert CLSM-CDF Backfill

The culvert is a corrugated steel horizontal ellipse (28'1" x 17'1"). This type of structure requires soil-structure interaction. The structure is required to carry H-20 or HS-20 live loads. The granular backfill is specified to be compacted to not less than 90 percent density per AASHTO T 180. Granular backfill at least 6 feet beyond the structures sides is required. The backfill must also extend to 4 feet above the structure. The culvert's length is 60 feet.

Table 3

Table from "Modulus of Soil Reaction Values For Buried Flexible Pipe", Howard, A. K. Reference 1.

Soil type-pipe bedding material (Unified Classification System[a]) (1)	E' for Degree of Compaction of Bedding, in pounds per square inch			
	Dumped (2)	Slight, <85% Proctor, <40% relative density (3)	Moderate, 85%–95% Proctor, 40%–70% relative density (4)	High, >95% Proctor, >70% relative density (5)
Fine-grained Soils (LL > 50)[b] Soils with medium to high plasticity CH, MH, CH-MH	No data available; consult a competent soils engineer; Otherwise use $E' = 0$			
Fine-grained Soils (LL < 50) Soils with medium to no plasticity CL, ML, ML-CL, with less than 25% coarse-grained particles	50	200	400	1,000
Fine-grained Soils (LL < 50) Soils with medium to no plasticity CL, ML, ML-CL, with more than 25% coarse-grained particles	100	400	1,000	2,000
Coarse-grained Soils with Fines GM, GC, SM, SC[c] contains more than 12% fines				
Coarse-grained Soils with Little or No Fines GW, GP, SW, SP[c] contains less than 12% fines	200	1,000	2,000	3,000
Crushed Rock	1,000	3,000	3,000	3,000
Accuracy in Terms of Percentage Deflection[d]	±2	±2	±1	±0.5

[a] ASTM Designation D-2487, USBR Designation E-3.

[b] LL = Liquid limit.

[c] Or any borderline soil beginning with one of these symbols (i.e., GM-GC, GC-SC).

[d] For ±1% accuracy and predicted deflection of 3%, actual deflection would be between 2% and 4%.

Note: Values applicable only for fills less than 50 ft (15 m). Table does not include any safety factor. For use in predicting initial deflections only, appropriate Deflection Lag Factor must be applied for long-term deflections. If bedding falls on the borderline between two compaction categories, select lower E' value or average the two values. Percentage Proctor based on laboratory maximum dry density from test standards using about 12,500 ft-lb/cu ft (598,000 J/m³) (ASTM D-698, AASHO T-99, USBR Designation E-11). 1 psi = 6.9 kN/m².

Table 4
Table from "E' and Its Variation With Depth", Hartley, J. D. and Duncan, J. M. Reference 5.

Type of soil (1)	Depth of cover (ft) (2)	Standard AASHTO Relative Compaction			
		85% (3)	90% (4)	95% (5)	100% (6)
Fine-grained soils with less than 25% sand content (CL, ML, CL-ML)	0–5	500	700	1,000	1,500
	5–10	600	1,000	1,400	2,000
	10–15	700	1,200	1,600	2,300
	15–20	800	1,300	1,800	2,600
Coarse-grained soils with fines (SM, SC)	0–5	600	1,000	1,200	1,900
	5–10	900	1,400	1,800	2,700
	10–15	1,000	1,500	2,100	3,200
	15–20	1,100	1,600	2,400	3,700
Coarse-grained soils with little or no fines (SP, SW, GP, GW)	0–5	700	1,000	1,600	2,500
	5–10	1,000	1,500	2,200	3,300
	10–15	1,050	1,600	2,400	3,600
	15–20	1,100	1,700	2,500	3,800

5.1 Engineering Requirements For The
    Granular Backfill

  - Compacted density, 90 percent.

  - 6 inch lifts to the bottomof
    the subbase.

  - Test each lift for proper
    compaction.

  - Place backfill 6 feet beyond
    each side of structure.

  - Place backfill 4 feet over top
    of structure.

5.2 Engineering Requirements For
    The CLSM-CDF Backfill

  - Meet or exceed the compaction
    requirement for granular
    backfill.

  - Place CLSM-CDF to the same
    dimensions for granular
    backfill; 6 feet beyond each
    side and 4 feet over top of
    structure.

  - Consider cost variations for
    backfill 3, 4, and 5 feet
    beyond outside of the
    structure.

  - No lift placement restrictions.

  - Regulate rate of placement to
    avoid flotation of culvert.

  - Make 4" x 8" compressive
    strength cylinders.

5.3 Quantities For Granular Backfill
    and CLSM-CDF

- Granular backfill = 1057 c.y.

- CLSM-CDF:
  Outside of Structure (c.y.)

Dist. =	6'	5'	4'	3'
CY =	1057	964	871	776

5.4 Cost Analysis

<u>Granular Backfill Air Tamped</u>

Quant.	Labor/	Material/	Total
cu.yd.	cu.yd.	cu.yd.	Cost
1057	$14.00	$7.00	$ 22,197

Testing per 6" Lifts
($ 250/day.estimate 3 days)   750

Total Cost Granular
Backfill = $ 22,947

<u>CLSM-CDF Backfill</u>

Quant.	Labor/	Material/	Total
cu.yd.	cu.yd.	cu.yd.	Cost

CLSM-CDF 6' Outside Structure

| 1057 | — | $25.00 | $ 26,425 |

CLSM-CDF 5' Outside Structure

| 964 | — | $25.00 | $ 24,100 |

CLSM-CDF 4' Outside Structure

| 871 | — | $25.00 | $ 21,775 |

CLSM-CDF 3' Outside Structure

| 776 | — | $25.00 | $ 19,400 |

Testing
(Flat fee for 2 cylinders) $ 100

Total Cost CLSM-CDF backfill,
depending on trench width =

        $ 20,400 to $ 27,425

   In this case the cost of CLSM-CDF,
if backfilled to the same dimensions
and not considering any of the
previously referenced construction
items, would cost more than
conventional backfill. Reduced
horizontal widths, because of the
flowability of CLSM-CDF, would
result in lower costs for the
CLSM-CDF backfill.

6 FUTURE RESEARCH

The Ohio Department of
Transportation (ODOT) is currently
funding research for the use of
CLSM-CDF in the construction of a
pipe arch (Paulding County, Ohio)
and a long span arch (Cuyahoga
County, Ohio). The pipe arch was
built in 1989 and the long span arch
is scheduled for construction in
1990. Research reports for both
structures will be completed in
1991. Interim reports on materials
and CLSM mixture designs are

currently being completed.

Future research holds the keys to CLSM's eventual potential of acceptance and use. Research is needed in all areas; uses, materials, testing manufacturing, specifications, and economics.

The research of CLSM components needs to consider <u>all</u> industry generated waste or byproducts. The use of CLSM provides an excellent opportunity for the use of a number of industry waste or by products in addition to fly ash. These materials must be researched to be sure that they will meet CLSM use requirements. Their use, in CLSM mixtures, should also consider their environmental effects.

Specialized mixtures need to be studied. These specialized mixtures include mixtures for; corrosion, structural, pavements, and thermal uses.

7 Summary

Controlled Low Srength Materials (CLSM) are being sold throughout the world by a variety of names. Technical information is still needed for the many possible applications. CLSM-CDF for the backfill around flexible structures has yielded data that shows that the material would reduce the horizontal deflection by 1/2 when compared to conventional backfill for the same modulus of soil reaction (E') values.

The modulus of soil reaction (E') can be determined for CLSM-CDF mixtures when Young's modulus (Es) and the Poisson ratio (Vs) are known. Laboratory data presented shows correlation between CLSM-CDF's compressive strength and Young's modulus (Es). Calculations were made according to published findings for modulus of soil reaction (E)' using Young's modulus (Es) and the Poisson ratio (Vs).

Trial mixes for CLSM-CDF should be made for untested materials to determine flowability and compressive strength. Standard design mixtures should result in good flow characteristics as

measured by a flow cylinder. If removability is considered, the 28 day compressive strength should be 100 psi or less.

CLSM-CDF backfill can be cost effective for the backfilling around flexible structures. It offers additional savings when considering construction time, safety, and reliability of the backfill. Possible material suppliers for CLSM-CDF need to review cost control methods. Unit costs for CLSM-CDF should be in the range of 50% of the cost of a 600 pound portland cement mixture.

References

1. "Modulus of Soil Reaction Values For Buried Flexible Pipe", Howard, A. K., <u>Journal Of The Geotechnical Engineering Division</u>, ASCE, January 1977.

2. <u>Soil Engineering</u> (Third Edition), Spangler, M. G. and Handy, R. L., Intext Educational Publishers, New York, 1973.

3. "Buried Structures", Watkins, R. K., <u>Foundation Engineering Handbook</u>, Winterkorn, H. F. and Fang, H. Y., Van Nostrand Reinhold Company, New York, 1975.

4. <u>Design and Construction of Sanitary and Storm Sewers</u>, WPCF Manual of Practice No. 9, Washington, D.C., 1970.

5. "E' and Its Variation With Depth", Hartley, J. D. and Duncan, J. M., <u>Journal of Transportation Engineering</u>, ASCE, Sept. 1987.

6. <u>HANDBOOK OF STEEL DRAINAGE & HIGHWAY CONSTRUCTION PRODUCTS</u> (2nd Edition), American Iron and Steel Institute, New York, 1971.

7. <u>Hardened Concrete: Physical and Mechanical Aspects</u>, Neville, A. M., American Concrete Institute, Detroit, Michigan, 1971.

8. "Soil Technology And Engineering Properties Of Soils", <u>Foundation Engineering Handbook</u>, Winterkorn, H. F. and Fang, H. Y., Van Nostrand Reinhold Company, New York, 1975.

Structural Performance of Flexible Pipes, Sargand, Mitchell & Hurd (eds) © 1990 Balkema, Rotterdam. ISBN 90 6191 165 6

Structural stability of deflected flexible pipe in a backfilled trench

L. H. Daniels
Consulting Engineer, Novato, Calif., USA

ABSTRACT: The often specified vertical deflection limit of 5% of diameter implies that pipe with greater deflection may not be structurally stable. The paper addresses the question of what load can such a pipe safely carry. Allowable depth of cover for flexible pipe is often several times greater than actual conditions and a pipe ellipsed as much as 15% may be a safe and serviceable structure.

The specified deflection limits are needed as a measure of backfill workmanship. However, when the problem is to judge structural stability of an improperly backfilled flexible pipe, deflection limits are meaningless. If the installed pipe has smooth curves free of kinks or buckles, strength can be conservatively evaluated by an analogy with pipe of a larger diameter. Formulas and charts are presented as an aid to evaluation for steel pipe.

1 INTRODUCTION

The specification limit for ring deflection of buried flexible pipe is sometimes exceeded when the pipe installation method does not result in suitable backfill density. This constitutes "bad" workmanship and can be grounds for rejection. However, no one really wants to reject and cause removal of an installed pipe unless it is actually functionally defective. The specified maximum ring deflection limit may or may not be a good indicator of serviceability. Ring deflection is defined as the ratio of decrease in vertical diameter to original diameter.

Figure 1

Certainly deflection is a good indicator of backfill quality, and perhaps is a more relevant inspection parameter than backfill density measurements. Figure 1 shows the use of a tubular story pole that makes routine pipe shape measurements easy. If the contractor monitors pipe shape as the job progresses, then the backfilling methods can be controlled to keep deflections within limits. Larger ring deflection is presumed to be caused by consolidation of low density backfill below the haunch and in the side fill. Ring deflection greater than 2% of the diameter is an indication that backfill methods are not achieving suitable density. The historical limit for deflection is 5%.

The ability of flexible pipe to move into and thus consolidate poorly compacted backfill is very valuable. Equilibrium is established and the resulting pipe shape, subject to evaluation, may be safe and serviceable. The same poorly compacted backfill around rigid pipe may not be detected during construction, but may be the cause of future problems, such as pavement damage or pipe disjointing.

If an economically significant length of pipe has been installed before the excessive deflection is discovered, strength of the installed pipe as affected by actual shape may become an acceptance criteria.

Deflection consists of two parts, initial deflection when the backfill is brought up to grade and rapidly declining continuing deflection due to soil consolidation. The time between completion of backfill and construction of roadway base and paving is usually long enough to protect the pavement against further deflection. When excessive deflection has taken place, the pipe is in equilibrium with the soil load and surrounding envelope, but the factor of safety against collapse is unknown.

Published tables of allowable height of cover typically use a factor of safety of two and may include a load factor based upon backfill density. Highway culverts are sometimes placed under deep fills where ring compression strength becomes limiting. In storm sewer work, soil cover is usually small enough that strength is not limiting. Actual height of cover seldom approaches the maximum allowable as limited by ring compression strength. As depth of cover becomes smaller the factor of safety becomes larger.

Under very shallow cover deflection, due to live load, must be liimted to protect the pavement, but that is another topic. The purpose of this paper is to set forth a method by which the structural stability of flexible pipes that have deflected under the loads imposed by backfill can be evaluated.

2 ASSUMPTIONS

Development of a clean simple method of analysis depends upon a properly constructed set of assumptions. This is where judgment is applied to the problem and rules of the game are developed. The assumptions that follow also are the limits of validity of the method of analysis based upon them.

1. The deflected pipe shape consists of smooth curves that have resulted from uniform bedding conditions. Unsymetrical shapes, hinges, kinks, or severe local buckles caused by unbalanced backfill or construction equipment will not be considered here.

2. The shape of the crown and invert can be approximated by a circular arc with a radius larger than the radius of the round pipe.

3. Ring compression in the crown and invert arcs is comparable in magnitude to ring compression in a round pipe of the same radius under the same fill height.

4. Ring deflection has stabilized and shows no significant change with time. This implies good in-situ soils or wide enough side fill to preclude continued plastic deformation of soft insitu soils outside of the trench.

3 STRENGTH MODEL

Figure 2 illustrates the relation between a deflected pipe and a hypothetical pipe of larger diameter. The actual shape of the deflected pipes, that I have observed, appears to be elliptical. However, no confirming measurements have been made.

The expression for the diameter of the assumed equivalent pipe was derived from rules for constructing an ellipse from circular arcs using 3 radii. No claim is made as to exactness. Convenience is important and only two easily made measurements are required, the horizontal diameter and the vertical diameter.

The diameter of the hypothetical pipe can also be determined by measurement of the mid-ordinate from a chord of suitable length. In figure 2, as the ellipse approaches a circle, the chord of the top arc approaches seven tenths of the diameter of the original pipe. This seems to be a good choice for a chord from which to measure the mid-ordinate.

Figure 3 provides the elementary relationship between chord length, mid-ordinate and diameter.

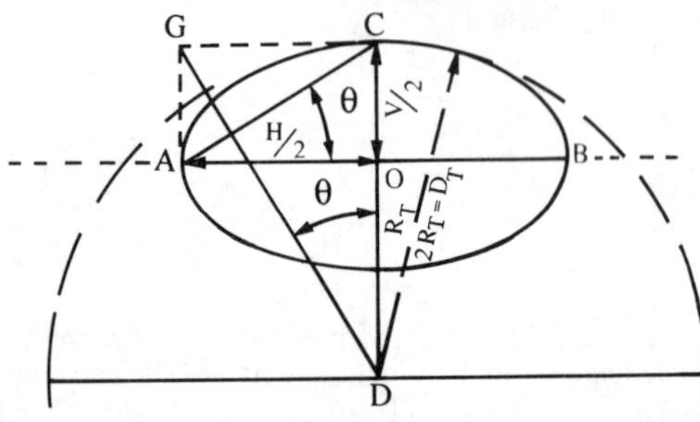

CENTER D IS FOUND BY LINE FROM $G\perp$ to AC
FROM TRIANGLES OAC&DCG

$$\frac{R_T}{H/2} = \frac{H/2}{V/2} \qquad d = \%\ \text{Vert. defl.}$$

$$\frac{D_T}{H} = \frac{H}{V} \qquad i = \%\ \text{Horiz. defl.}$$

$$\frac{D_T}{H} = \frac{H^2}{V} \qquad \frac{D_T}{D_O} = \frac{(1+i)^2}{(1-d)}$$

Figure 2

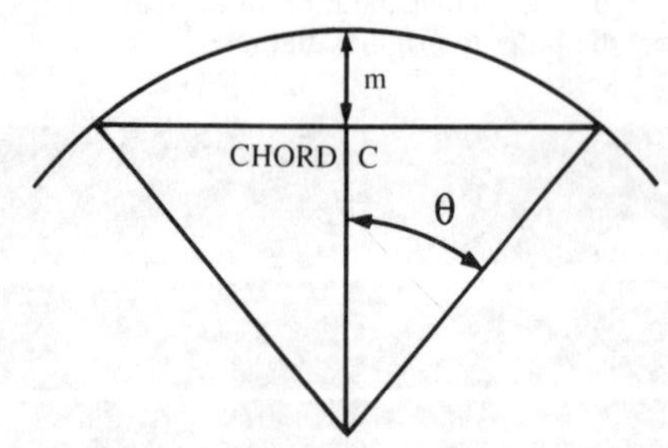

MEASURE M FROM CHORD OF LENGTH C

$$\theta = 2\ \text{ARCTAN}\ \frac{2m}{C}$$

$$R_T = \frac{C}{2\,\text{Sin}\theta}$$

$$D_T = \frac{C}{\text{sin}\theta}$$

Figure 3

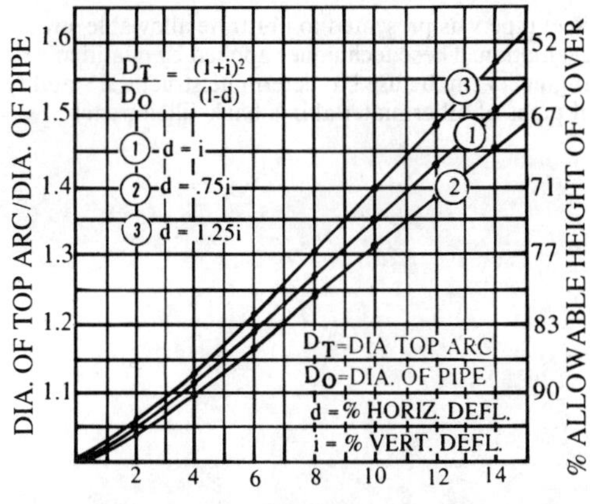

PROPERTIES OF ELLIPSED PIPE

Figure 4

If an assumption is made as to the relationship between the percent decrease in vertical diameter and the percent increase in horizontal diameter, values for the diameter of the hypothetical pipe can be obtained from the expression developed in figure 2. Figure 4 presents the results of three possibilities in dimensionless form. The most likely possibility is that the increase in horizontal diameter "i" is approximately equal to the decrease in vertical diameter 'd'. Certainly one would not expect ratios greater than d=1.75i, nor less than d=0.75i, the other two assumptions plotted. For deflections below 8% there is little difference between assumptions. In practice, the actual ratio would be known from measurements.

The vertical scale is shown in two forms. The ratio of the diameter of the top arc to the diameter of the pipe is obtained from the expression in the figure. Since soil load or thrust in the pipe wall is directly proportional to pipe diameter, the decrease in allowable height of cover is the inverse of the ratio of the diameter of the hypothetical pipe to the original diameter.

For example: a 60″ 14 ga spiral rib pipe would have an allowable fill height of 29 feet, based on design requirements for thrust. If it deflects 8% after trench backfill has been placed, the allowable cover would be 80% of 29 feet or 23 feet. Pipe in storm drain projects are seldom that deep below the surface.

The deflected pipe is safely in the ground and in equilibrium with imposed loads. Certainly handling stiffness is not limiting. However, allowable cover percentages in figure 4 do not consider the increased possibility of wall buckling due to the flatter curves of the hypothetical pipe. Stress in the hypothetical pipe should be checked against allowable stress.

4 ALLOWABLE STRESS

Allowable stresses establish a safe approach to design in the face of variable field conditions, while the Engineer evaluating an inplace structure is able to observe actual conditions. Most often safety factors will be discovered to be greater than 2 but where fills are deep an understanding of load factors and safety factors can be important.

The allowable stress formulas in the handbook of steel drainage and highway construction products were developed from physical testing rather than theoretical analysis. Bending stresses that result from the outward movement of the springline do not effect the ultimate ring compression capacity. The handbook sets forth allowable stresses as follows:

$$\text{For } \frac{D}{r} < 294, \ fa = \frac{33,000}{2K}$$

$$\text{For } \frac{D}{r} > 294 \ \& < 500, \ fa = \left[40,000 - .081\left(\frac{D}{r}\right)^2\right] \div 2K$$

$$\text{For } \frac{D}{r} > 500, \ fa = \frac{4.93 \times 10^9}{2K\left(\dfrac{D}{r}\right)^2}$$

Where fa = Allowable Wall Stress, psi
       K = Load Factor
       r = Radius Gyration, in.
       D = Diameter, in.

For 85% Standard Density, K = 0.86
For 90% Standard Density, K = 0.75
For 95% Standard Density, K = 0.65

The divisor 2K in each formula consists of the safety factor 2 and the load factor K that has resulted from research conducted at Utah State University. The load factor is related to the composite nature of the pipe and soil. A load factor of 1 results when all of the backfill load is carried by the pipe. The research on buried corrugated steel pipes resulted in the lower values of K.

The relatively high safety factor of two is used to account for the variability of field conditions as compared to test conditions. If field conditions are known and equilibrium has been established, the safety factor of two is perhaps higher than is justified.

The interaction between the soil and the pipe is affected by pipe deflection. As the pipe crown moves downward the soil arch must follow and while following tends to take an increased share of the load. The side fill is consolidated by the outward movement of the springline and must buttress the horizontal thrust at the springline as well as support the reaction of the soil arch. Since consolidation of the side fill is a time related phenomena, stability should be observed over a significant time period, perhaps 30 days.

If K were chosen as 0.75 with a safety factor of 2. the product becomes 1.5. If the value for the product were retained at 1.5 and K were increased to 1 the safety factor becomes 1.5. A load factor of I results when all of the backfill load is carried by the pipe.

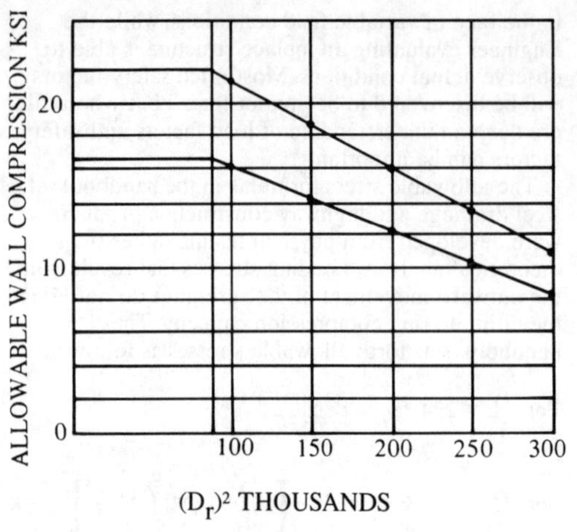

(D_r)² THOUSANDS

Figure 5

On the other hand ASTM Specifications do not discuss load factors, but do use a safety factor of two. A load factor of one is implied which seems conservative for deflected pipe.

Figure 5 is the graphical presentation of the allowable stress formulas with the product of safety factor times the load factor equal to 1.5 and 2.0.

Example

A 90 inch diameter 12 gauge spiral rib pipe with 12 feet of cover has deflected 8 percent. What is the level of concern and course of action recommended?

From figure 4 with d=i Dt/Do =1.27. This indicates an equivalent diameter of 115″.

D/r=115/.355=324

and (D/r)² in thousands is 105.

From figure 5 using the more liberal curve allowable wall compressive stress is 21 KSI. The effective area of 0.883 sq. in./ft. indicates an allowable wall compression of 18,543 lbs. per ft. and an allowable height of cover of 41 ft. The use of a safety factor times load factor of two would reduce the allowable cover to 30 feet.

It appears that the pipe with only 12 feet of cover could be left in place, but an end to end inspection should be undertaken. Smooth curves with no hinges, kinks, or severe local buckles would be required and deflection should be monitored for change with time.

5 CONCLUSION

A methodology has been presented for evaluation of deflected flexible pipe with smooth curves in stable equilibrium with imposed soil load. A stress check for steel pipe was presented to illustrate allowable stress evaluation. These techniques applied by qualified engineers can be used to determine structural stability of pipes of other material in a back-filled trench.

122

Structural Performance of Flexible Pipes, Sargand, Mitchell & Hurd (eds) © 1990 Balkema, Rotterdam. ISBN 90 6191 165 6

Nonlinear analysis of raising of flexible pipe strings

Ronald S. Reagan
University of Southern Mississippi, Long Beach, Miss., USA

ABSTRACT: Experimental results for lifting an 80 foot length of 1 inch diameter pipe are presented. The free end of the pipe string was first raised with a vertical force to a height of 30 ft. Then with the pipe end held at the 30 ft elevation, the free end was moved back over the unsuspended portion of the string until the rotation at the top of the pipe was 108 deg. Yielding of the pipe occurred during the experiment.

A nonlinear two-point boundary value problem was posed which models the pipe raising experiment. The ordinary differential equations, boundary conditions, foundation model, and moment-curvature relation for the model are presented. The boundary value problem was solved using the multiple shooting algorithm. The computed results for the lifting portion of the experiment agree very closely with the experiment. For the moving back portion the computed resultant pulling force is low by about 10%.

1 PURPOSE

The purpose is to 1) present the results of an experiment in which a pipe string was lifted from its free end and 2) describe a nonlinear analysis of the mechanics problem posed by the experiment. The experiment was conducted by Carl G. Langner (unpublished results) of Shell Development Company. The problem makes a useful test case for pipe handling analysis software because it includes both material and geometric nonlinearities.

2 LIFTING EXPERIMENT

In this experiment an 80 ft length of 1-in. nominal diameter steel pipe was first laid out horizontally on a plywood deck. The pipe was weighted with lead weights to give it a density heavier than seawater so as to represent a submerged offshore pipeline. One end of the pipe was clamped to the deck. The setup is shown in Figure 1.

The lifting experiment consisted of two phases. In the first phase the free end was raised to a height of 30 ft by a vertical force. This is typical of offshore pipeline operations. In the second phase the raised end was held at the 30 ft height and moved horizontally to

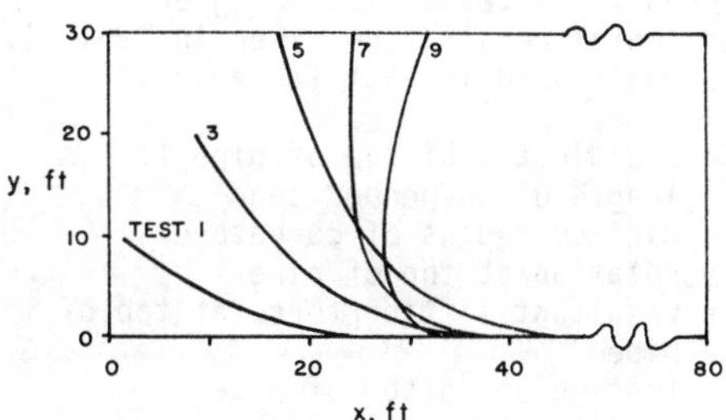

Figure 1. Experimental Setup

Figure 2. Calculated Configurations

bend the pipe back over the pipe remaining on the deck. The deflected shapes (analysis results) plotted in Figure 2 show the lifting sequence.

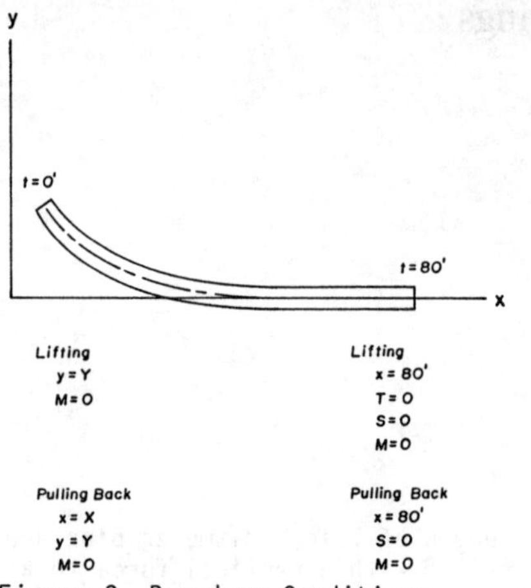

Figure 3. Boundary Conditions

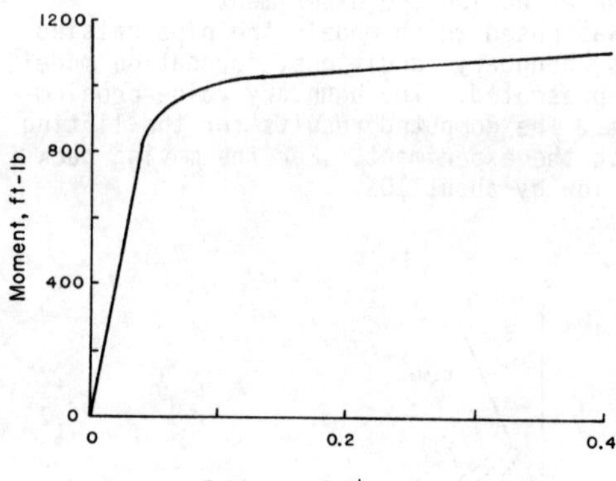

Figure 4. Moment-Curvature Diagram

Configurations measured during the raising were designated as tests 1-5, and those observed during the pulling back are referred to as tests 6-9. Langner's experimental results are given in Table 1. The symbols used in that Table are:

Y,X = coordinates at top of pipe
L = length of suspended span
R = minimum radius of curvature
a = rotation at top of pipe
F = resultant lifting force at top of pipe
T_o = tension at bottom of pipe
T^o = tension at top of pipe
S = shear at top of pipe

Note that the tension in the pipe is zero for the lifting phase, while portions of the pipe are in compression for the pulled back configurations.

3 MATHEMATICAL MODEL

The mathematical model can be expressed as a nonlinear two-point boundary value problem. The components of the model are ordinary differential equations, boundary conditions at the ends, a foundation support relation, and a constitutive model for the pipe.

The ordinary differential equations for the problem as given by Love (1944) are written as Equations 1-6.

$$x' = \cos a \tag{1}$$
$$y' = \sin a \tag{2}$$
$$a' = K \tag{3}$$
$$T' = KS - (w-f)\sin a \tag{4}$$
$$S' = -KT - (w-f)\cos a \tag{5}$$
$$M' = -S \tag{6}$$

where $' = d/dt$
t = arc length (measured from pulled end)
x,y = coordinates (see Figure 1)
a = rotation angle
T = tension
S = shear
M = bending moment
K = curvature
w = weight per unit length
f = elastic foundation reaction
$= 0$ for $y > 0$
$= ky$ for $y < 0$

Equations 1-3 enforce compatibility of the displacements, and Equations 4-6 are the equations of equilibrium.

The boundary conditions imposed at the ends of the pipe for the lifting and pulling back phases are shown in Figure 3.

The support of the pipe by the deck was modelled by a one-way Winkler foundation. The spring constant $k = 47$ lb/ft² was chosen so that the pipe would sink 0.1 ft into the foundation under its own weight of 4.7 lb/ft. This model was stiff enough to represent the actual support conditions and soft enough to avoid numerical difficulties in the solution.

The nonlinear moment-curvature relationship was taken from Langner (unpublished result) as Eq. 7.

$$K = M/EI + A (M/M^*) \, abs \, (M/M^*)^B \tag{7}$$

The values of the constants appearing in Equation 7 are $EI_* = 21710$ lb-ft², $A = 1.160$ ft-1, $M^* = 1178$ ft-lb, and $B = 19$. They are based on an outside pipe diameter of 1.315 in, a wall thickness of 0.175 in, and a yield strength of 56.8 ksi. The resulting moment-curvature diagram appears in Figure 4.

Table 1. Experimental data from suspended pipe span tests (1-inch schedule 80 pipe, lead weighted to 4.7 lb/ft)

Test	Y(ft)	X(ft)	L(ft)	R(ft)	a(deg)	F(lb)	T_0(lb)	T(lb)	S(lb)
1	10	2.4	34	37.5	35.0	80	0	45.9	66.5
2	15	4.9	38	30.9	47.5	96	0	70.8	64.9
3	20	8.1	42	26.9	57.7	110	0	93.0	58.8
4	25	12.1	46	24.0	67.0	128	0	117.8	50.0
5	30	16.4	50	23.6	73.3	148	0	141.8	42.5
6	30	19.3	46	13.8	82.7	135	-20	129.9	36.8
7	30	23.8	45	7.06	91.7	124	-32	120.7	28.5
8	30	27.2	43	6.54	99.7	110	-41	107.5	23.4
9	30	31.3	43	5.34	107.8	104	-45	103.0	14.1

Table 2. Calculated results corresponding to experimental data

Test	Y(ft)	X(ft)	L(ft)	R(ft)	a(deg)	F(lb)	T_0(lb)	T(lb)	S(lb)
1	10	2.4	32.9	36.4	35.4	81.5	0	47.2	66.4
2	15	4.9	38.3	30.4	47.6	95.7	0	70.7	64.6
3	20	8.1	42.1	27.0	58.2	110.8	0	94.2	58.4
4	25	12.1	46.3	24.7	67.3	127.6	0	117.7	49.2
5	30	16.6	50.6	23.1	74.6	146.5	0	141.2	38.8
6	30	19.3	47.3	14.8	83.3	122.1	-25.6	115.6	39.4
7	30	23.8	44.3	8.40	94.1	105.3	-41.4	99.6	34.4
8	30	27.2	43.1	6.16	100.9	99.1	-46.3	94.9	28.8
9	30	31.3	42.5	4.64	108.0	94.6	-49.0	92.1	21.6

4 SOLUTION PROCEDURE

A multiple shooting method was used to solve the boundary value problem for each of the configurations reported for the experiment. Langner (unpublished result) had reported that his finite-difference approach failed for the pulling back configurations when the pipe was in compression. The FORTRAN program MSHOOT obtained from Sandia Laboratories (Scott & Watts, 1976) was adapted to solve the problem. MSHOOT combines other numerical routines (integrators of ordinary differential equations, a linear equation solver, and a nonlinear equation solver) to provide a solver for nonlinear boundary value problems. The MSHOOT user must provide routines to evaluate the ordinary differential equations and the boundary conditions.

In multiple shooting the boundary value problem is reduced to a set of nonlinear functions that need to be made zero. The independent variables are taken as the boundary values at one end of the problem plus the values of the dependent variables in the differential equations at each of the interior shooting points. The dependent variables (nonlinear functions) are taken as the errors (residuals) in the boundary conditions at the ends plus the differences in the dependent variables on the opposite sides of each shooting point. Numerical integration is used to evaluate the nonlinear functions. The independent variables are adjusted by a nonlinear equation solver until a convergence tolerance is met.

Increasing the number of shooting points improves the robustness of the method while also increasing the computational time. Three equally spaced shooting points were used after some experimentation.

The analysis results are given in Table 2 in the same form as the experimental results. For the lifting part of the problem the experimental and analytical results agree closely. However for the pulling back portion the calculated pulling forces are about 10% high. This is thought to be due to the inaccuracy of the moment-curvature model. The inclusion of a more accurate unloading rule could improve the results.

Further analysis results are shown in Figure 5 which shows a plot of the pulling force versus the vertical displacement for the lifting phase. Deflected shapes for Test 9 as measured in the experiment and as calculated are plotted for comparison in Figure 6.

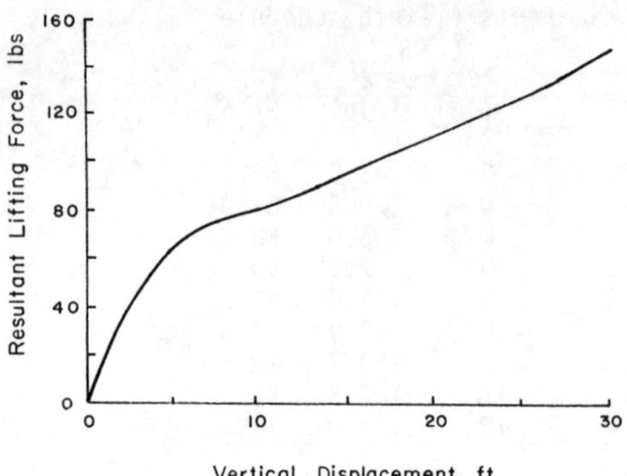

Figure 5. Resultant Lifting Force vs. Vertical Displacement

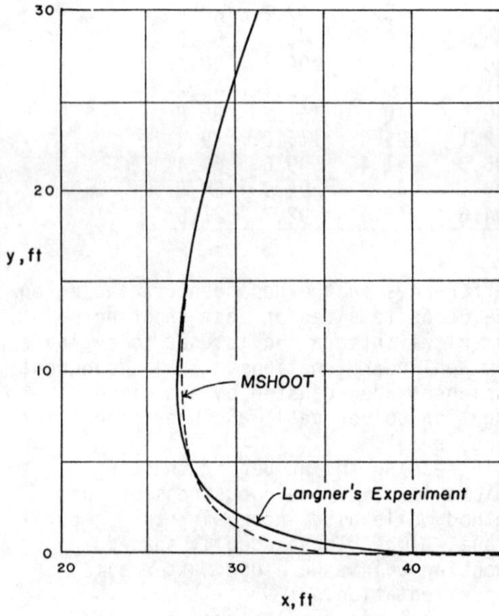

Figure 6. Experimental and Analytical Configurations for Test 9

5 CONCLUSIONS

The multiple shooting method used herein provides a solution to the posed pipe lifting problem which is accurate enough for most purposes. The problem considered is a good test problem for nonlinear analysis since it exhibits both large rotations (108 degrees) and yielding of the pipe.

ACKNOWLEDGEMENT

The author would like to thank Carl G. Langner of Shell Development Company for permission to use the results for the pipe lifting experiment.

REFERENCES

Love, A. E. H. 1944. A Treatise on the Mathematical Theory of Elasticity. p. 381-413. New York: Dover.
Scott, M. R. and Watts, H. A. 1976, "A Systematized Collection of Codes for Solving Two-Point Boundary-Value Problems", Numerical Methods for Differential Systems, Edited by L. Lapidus and W. E. Schiesser, Academic Press, N. Y., pp. 197-227.

Structural Performance of Flexible Pipes, Sargand, Mitchell & Hurd (eds) © 1990 Balkema, Rotterdam. ISBN 90 6191 165 6

Pipeline reconstruction without excavation

R.D. Rice
Insituform of North America, Inc., Memphis, Tenn., USA

ABSTRACT: Over the past decade, in-place pipeline rehabilitation has been one of the fastest growing segments of infrastructure repair. Several innovative technologies emerged, all of which create a new thermoplastic or thermosetting plastic pipe-within-a-pipe. Additional advances and refinements in trenchless reconstruction now permit rebuilding of gravity pipe systems and some pressure pipe systems without excavation or with less excavation than total replacement. Through the use of cured-in-place pipe, the fold-insert-and-expand process and sliplining, infiltration can be controlled while restoring structural integrity to deteriorated, leaking systems. These advancements include techniques for reconstructing laterals from inside mains to and beyond the property lines, structural rebuilding of existing manholes, and a means of introducing solid wall plastic pipe into sewers without excavation and without leaving an annular space between old and new pipe walls.

1 INTRODUCTION

Increasing awareness of deteriorating infrastructure systems has led to new pipeline rehabilitation methods. However, the emergence of various technologies for trenchless and semi-trenchless pipe rehabilitation has caused considerable confusion as to appropriate application, performance characteristics, structural integrity, installation techniques, and true cost over the useful life of the product.

The intent of these various technologies is to provide a nondisruptive means for rebuilding our decaying underground piping systems. For ease of clarification, new and existing technologies can be grouped under four broad categories:

1. Cured-In-Place Pipe (CIPP)
2. Fold and Formed
3. Sliplining
4. Open Cut

Only processes which leave a structural pipe within the old existing pipe will be considered. Of these four types of construction, only two offer pipe replacement without excavation, cured-in-place pipe and fold and formed. Partial excavation may be required for sliplining, although it could be considered "trenchless" or semi-trenchless construction.

Pipeline Reconstruction Process Categories

NO EXCAVATION:

Cured-In-Place Pipe
Fold and Formed

EXCAVATION REQUIRED:

Sliplining
Open Cut

2 CURED-IN-PLACE PIPE

The CIPP process uses thermosetting resin technology to form a new pipe directly against the walls of the old, deteriorated host pipe, using the existing pipe as a mold or form. There are four different installation methods for CIPP:

1. Inverted in place with water
2. Inverted in place with air
3. Winched in place, water inflated
4. Winched in place, air inflated

All four CIPP processes follow the same general overall steps. First, the line to be rehabilitated is televised with portable closed circuit television equipment. This process determines the length and diameter of the line and the location of active and inactive service laterals. The line is then thoroughly cleaned and then rehabilitated with the CIPP process. The line is then again televised to ensure proper installation, and the laterals are reinstated, usually from the inside.

The most successful and widely used CIPP process is inverted in place with water. (See Figure 1) It has been used in the United States for over 12 years and 19 years worldwide. This proven technology has been used to reconstruct pipelines from 4" to 108" in diameter. Over 8 million feet have been installed in the United States and over 11 million worldwide. Cured-in-place pipe - water inverted (CIPP-WI) uses a nonwoven polyester felt tube, coated on the outside with a tough plastic layer. This tube is manufactured to the diameter and length required by the pipeline to be reconstructed and is installed by trained crews. The wall thickness is determined by standard buried flexible pipeline equations and can be increased by simply adding more layers of polyester felt. Once the old pipeline is cleaned, this CIPP tube is vacuum impregnated with a liquid thermosetting resin and is hydraulically inserted into the existing pipeline by the inversion method. Water used to invert the tube is then heated, initiating a cure which transforms the pliable tube into a hard and structurally sound pipe. The reinstatement of service laterals, in small diameter pipe, is accomplished by a remote controlled robotic cutter.

Installation by this process is covered by ASTM F-1216. Structural integrity has been validated through testing by Utah State University (1) and other organizations.

Processes which use the inversion technique for insertion are the most versatile and least traumatic to the pipe being installed. The processes which are installed by tying a rope or cable to the tube and then dragging the tube into position and through the old deteriorated pipe can cause damage to the tube during installation.

CIPP processes which use the inversion technique for insertion as well as those that use it for inflation have an advantage in that any standing water within the existing host pipe is pushed out of the

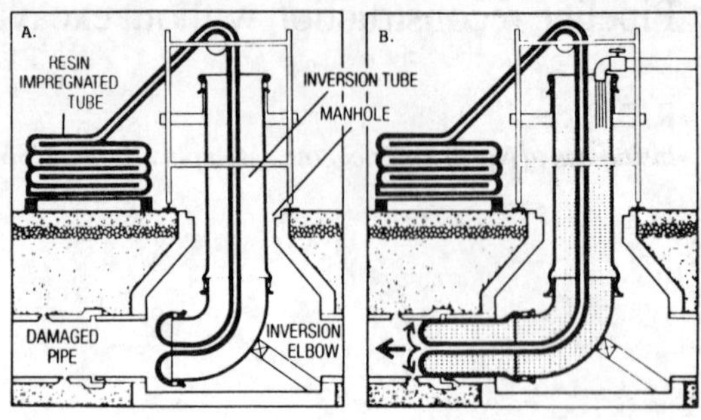

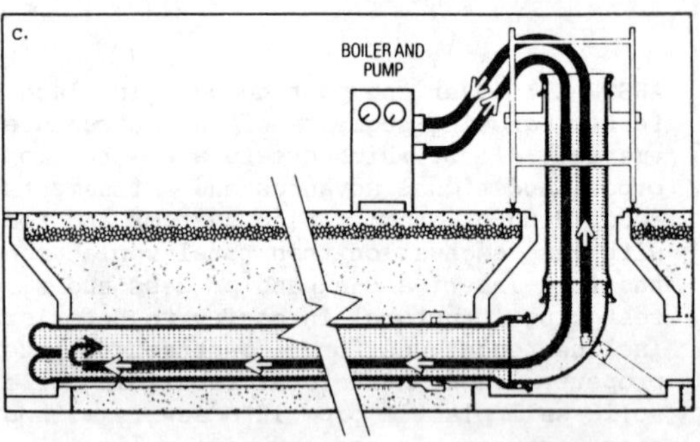

Figure 1.

A. A special needled felt reconstruction tube, coated on the outside, is custom engineered and manufactured to fit the damaged pipe exactly. It is impregnated with a liquid thermosetting resin and lowered into a manhole through an inversion tube. One end of the tube is firmly attached to the lower end of the inversion tube elbow.

B. The inversion tube is then filled with water. The weight of the water pushes the tube into the damaged pipe and turns it inside out, while pressing the resin impregnated side firmly against the inside walls of the old pipe. The smooth coated side of the tube becomes the new interior surface of the pipe.

C. After the tube is inverted through the old pipe to the desired length, the water is circulated through a boiler. The hot water causes the thermosetting resin to cure within a few hours, changing the pliable tube into a hard, structurally sound, pipe-within-a-pipe. It has no joints or seams and usually increases flow. The ends are cut off and the inversion tube and scaffolding are removed. Normally, there are no messy excavation repairs to be made since most work is done without digging or disruption.

way as the inverting face progresses
through the old deteriorated pipe. This
eliminates the possibility of trapped
water or air between the new CIPP pipe and
the old host pipe. Trapped water causes a
reduction in cross-sectional area as well
as weakening the structural capacity of
the new CIPP pipe. Processes which
inflate the tube like a balloon, causing
it to expand more or less uniformly from
end to end, do not have the unique ability
to push standing water out of the existing
pipe.

The soft tube used for the CIPP process
is made of either woven or nonwoven
fibers. In most cases, the fiber is
polyester or a similar material. The
purpose is to hold the resin in place
until it cures. Unless these fibers are
high-strength fibers such as Kevlar or
fiberglass, they do not add any
appreciable stiffness to the final pipe.
Polyester fibers are uniquely suitable for
this process as they hold resin quite well
and are highly resistant to chemicals and
gases found in sewer systems. Fiberglass,
on the other hand, can add strength to the
pipe wall, but is susceptible to attack
from acids commonly found in sewer
systems. Exposed fiberglass layers at
lateral openings and through porosity in
the wall are subject to corrosion and can
cause delamination, swelling and weakening
of the pipe wall.

Not all process types use vacuum impregnation. This is an important step in the
process as it eliminates trapped air
bubbles which cause porosity, which in
turn reduces wall strength and chemical
resistance. Vacuum impregnation pulls the
small trapped air bubbles out of the
resin, thereby giving a dense, solid and
homogeneous pipe wall.

Inverted in place tubes are preferred
because this installation technique causes
far less trauma during installation.
Water inversion is preferred over air,
again, because it is a more gentle insertion process. During the water inversion,
the wet-out tube floats within a water
medium, and since its specific gravity is
near that of water, there is very little
friction buildup during installation. On
the other hand, even though an air
inversion would be less traumatic than
winched in place, the uninverted portion
of the tube is dragged along the bottom,
against tube material already in place,
and can build considerable friction.

The strength of the finished cured-in-
place pipe depends almost entirely upon
the inherent characteristics of the
thermosetting resin used. Flexural

modulus characteristics of typical resins
used are in the range of 250,000 psi to
500,000 psi, flexural strengths 4,500 to
5,000 psi and tensile strengths 3,000 to
4,000 psi.

The thermosetting resins used in the
CIPP process have the unique ability to be
formulated to resist a wide variety of
reactive chemicals. Much of the current
quality assurance activity is targeted
towards the resin used to saturate the
felt tubing. The process may use polyester resins, epoxy resins, or vinyl ester
resins. The choice is an economic one
based upon the performance requirements of
a specific project. Polyester resins are
used more often because they have
strengths comparable to other resins, are
naturally resistant to acids, and are the
least expensive. However, certain
structural conditions or corrosive
environments may require the use of epoxy,
vinyl ester, or other special use resin.

In July 1988, a test was conducted at
Utah State University to determine the
structural support that can be provided to
a failing rigid pipe system through the
use of the Insituform® CIPP process. The
test was designed to determine the composite ring strength of the concrete/Insitu-
pipe™ system.

The test was conducted by placing two
parallel sections of pipe in the USU large
soil cell as shown in Figures 2 and 3.
The pipes in both test sections had been
broken prior to burial, and one section
then repaired using the Insituform
process. The two sections were separated
by a spacing of 7.5 feet center to center
and were 20 feet long. Height of the soil
cover over the tops of the test sections
was 3 feet. Bedding was a level plane of
compacted soil. The pipe zone backfill
soil (PZB) was silty sand placed in
12-inch layers. Vertical soil load was
applied by 50 hydraulic cylinders attached
to 10 reaction beams. Vertical and
horizontal diameters of both test sections
were measured after each increment of
load.

Vertical loads were then applied to the
soil cell in increments of 50 psi by the
hydraulic cylinders. Each increment of
loading applied a vertical soil pressure
at the top of the test pipes of 727 psf,
which is equivalent to 6.06 feet of soil
cover assuming a soil unit weight of 120
pcf.

Figure 4 provides a performance comparison of the two test sections based on a
load vs. deflection plot of the test data.
As shown in Figure 4, at a soil pressure
of approximately 3000 psf, the plots are

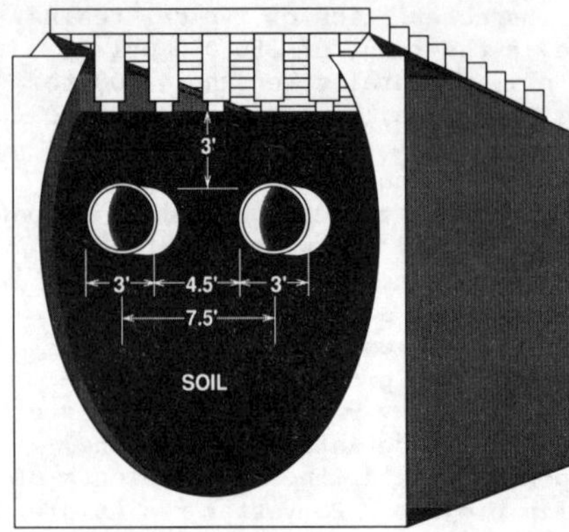

Figure 2. Cross section of the USU large soil cell showing the location of the test pipe sections in the Insituform test.

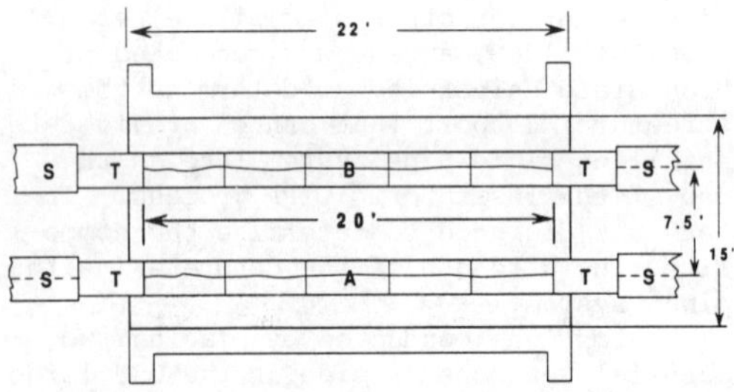

A - Insituform repaired pipe
B - Unrepaired pipe
S - Access pipe
T - Transition pipe

Figure 3. Plan view of soil cell with test pipes in place.

approximately linear. It should be noted that the slope of the Insituformed plot is not as steep as that of the un-Insituformed pipe, which means the Insituformed section deflected less than the unrepaired pipe. Table 1 provides additional test results.

From these tests, a comparison can be made between the theoretical analysis used for design by Insituform and the values obtained in these tests. The Insitupipe used in this test (30 inches by 21 millimeters with standard polyester resin properties) was designed to withstand a soil loading of 3600 psf (burial depth of 30 feet using soil density of 120 pcf). This design is controlled by buckling through the use of the Modified AWWA Equation. An upper load of 8724 psf (burial depth of 72.7 feet) of soil pressure was obtained in this test before any evidence of buckling failure occurred

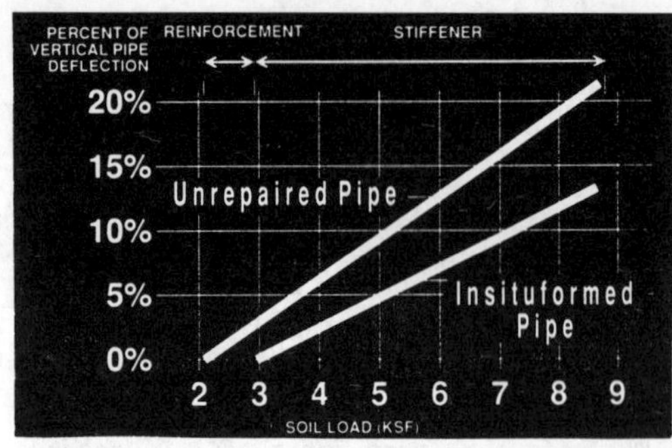

Figure 4. The reinforcement phenomenon is reflected by the amount of pressure necessary to begin pipe deflection. The stiffener phenomenon is reflected by the slope of the respective plots which shows the Insituformed pipe deflecting less than the unrepaired pipe.

Table 1.

Load psf	Equivalent Depth	Percent Deflection	
		Insituformed Pipe	Rigid Pipe
2181	18.2'	0	0.2
2908	24.2'	0	3.0
3635	30.3'	1.7	5.2
4362	36.4'	3.3	7.7
5089	42.4'	5.0	10.2
5816	48.5'	6.7	12.7
6543	54.5'	8.3	15.2
7270	60.6'	10.0	17.7
8724	72.7'	13.2	21.0

in the Insitupipe section. Using the design load (30 feet), it can be determined that the Insitupipe was designed with a minimum safety factor of 2.4 against buckling failure.

Design of flexible pipes may also be controlled through ring deflection. The generally accepted design deflection is 5%, which in itself has a minimum safety factor of 2.0. Even though the Insituformed section was designed for a maximum burial depth of 30 feet, the Insitupipe did not reach 5% deflection until an equivalent burial depth of 42 feet was reached.

This test showed that Insituform adds significant structural strength to buried, broken pipe. In addition to increasing the strength of cracked pipe, Insitupipe also retains service capability.

3 FOLD AND FORMED

The fold and formed process, which is available in 4" through 12" diameters in

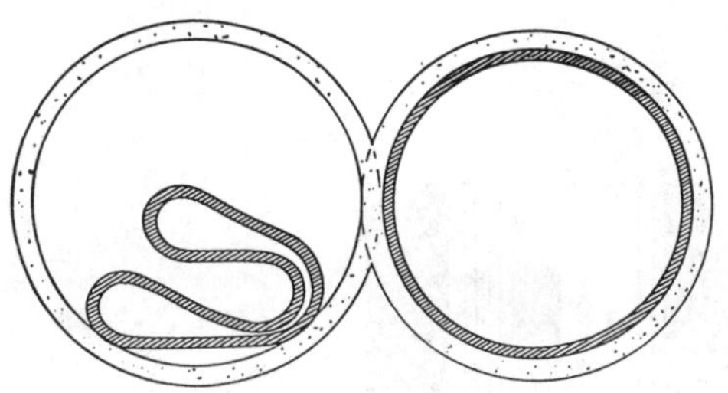

FOLDED CROSS SECTION   ROUNDED CROSS SECTION

Figure 5. Cross section of fold and
formed showing the folded configuration
(left) and the rounded configuration
(right).

PVC and 2" through 24" in HDPE, is a
unique process for reconstructing damaged
sewers and other pipeline systems in
municipal and industrial applications.
This reconstruction process involves the
insertion of a specially shaped plastic
extrusion (Figure 5) into an existing
pipeline, heating the plastic and then
expanding it against the interior surface
of the original conduit.

The specially formed plastic pipe is
fully extended within the existing conduit
and cut to length. Heat and internal
pressure are then applied, and through the
use of a specially designed rounding
device, for PVC, the plastic is rounded to
conform to the shape of the existing pipe
(Figure 6). For HDPE, only internal
pressure is used for expansion, hence the
pipe inflates simultaneously from end to
end rather than progressively.

This expansion produces dimples at the
side connections, making them easy to
locate using closed circuit television.
Pressure is maintained within the new
pipe, holding it against the original
conduit while the plastic is cooled to a
rigid state. To complete the installation, side connections are reinstated from
the inside of the line with a remote
controlled robotic cutting device. The
finished fold and formed pipe is
continuous from end to end with no seams,
no joints, no layers which could
delaminate, and adds some structural
strength to the entire rehabilitated line.

Even though the inside diameter of the
rehabilitated line has been reduced by the
wall thickness of the new pipe, the flow
capacity is generally not reduced.
Because the new pipe smooths out the
interior surface of the old line, bridging
over gaps, cracks and joints, and
considering plastic pipe's inherent
smoothness (Manning's "n" = 0.009 or 0.010

depending upon wall uniformity of original
pipe) (2,3,4), flow in any given line
segment is increased when repairing pipe
material such as concrete or vitrified
clay. Flow capacity would be reduced only
when the pipe to be repaired is a smooth,
nonporous, nonwetting material, such as
another plastic.

PVC and HDPE are widely used and
accepted thermoplastic pipe materials for
gravity and pressure lines for new installations. In the United States, approximately 90% of all municipal sewerage
systems use PVC pipe when the size
required falls within currently available
PVC pipe diameters. PVC and HDPE are
light in weight, have high impact
strength, are easily field cut and tapped,
have high chemical resistance and have
physical characteristics which make them
especially suitable for the fold and
formed installation process. However,
according to the American Society of Civil
Engineers and WPCF Manual of Practice on
Gravity Sanitary Sewer Design and Construction (5), PVC is the only thermoplastic
commonly used for gravity systems which is
not subject to environmental stress
cracking.

High density polyethylene is also used
in the fold and formed process. It is
installed in a similar manner but differs
in subtle but important ways. The HDPE
used in this process is subject to environmental stress cracking (ESC) and is
relatively weak compared to other plastics
which are typically used for piping
systems. ESC is a surface initiated,
purely physical phenomenon brought about
under polyaxial stress by an external
agent (i.e. wetting agents such as deter-
gents) which has no other discernible
effect on the polyethylene. These polyaxial stresses can be caused by external
stress such as ground water pressure and
thermal contraction of the restrained
pipe, as well as internal stresses caused
by the installation process and thermal
history (6). The expanding or forming
process currently used with HDPE inflates
the pipe more or less uniformly from end
to end and tends to trap any standing
water in the annular space.

Since PVC is not subject to environmental stress cracking, the expected life
of a PVC fold and formed system would be
the same as a newly installed PVC pipe.
HDPE, on the other hand, can have a short-
ened life expectancy in a highly stressed
or strained condition, especially in the
presence of wetting agents which acceler-
ate environmental stress cracking.
Although fold and formed technology is

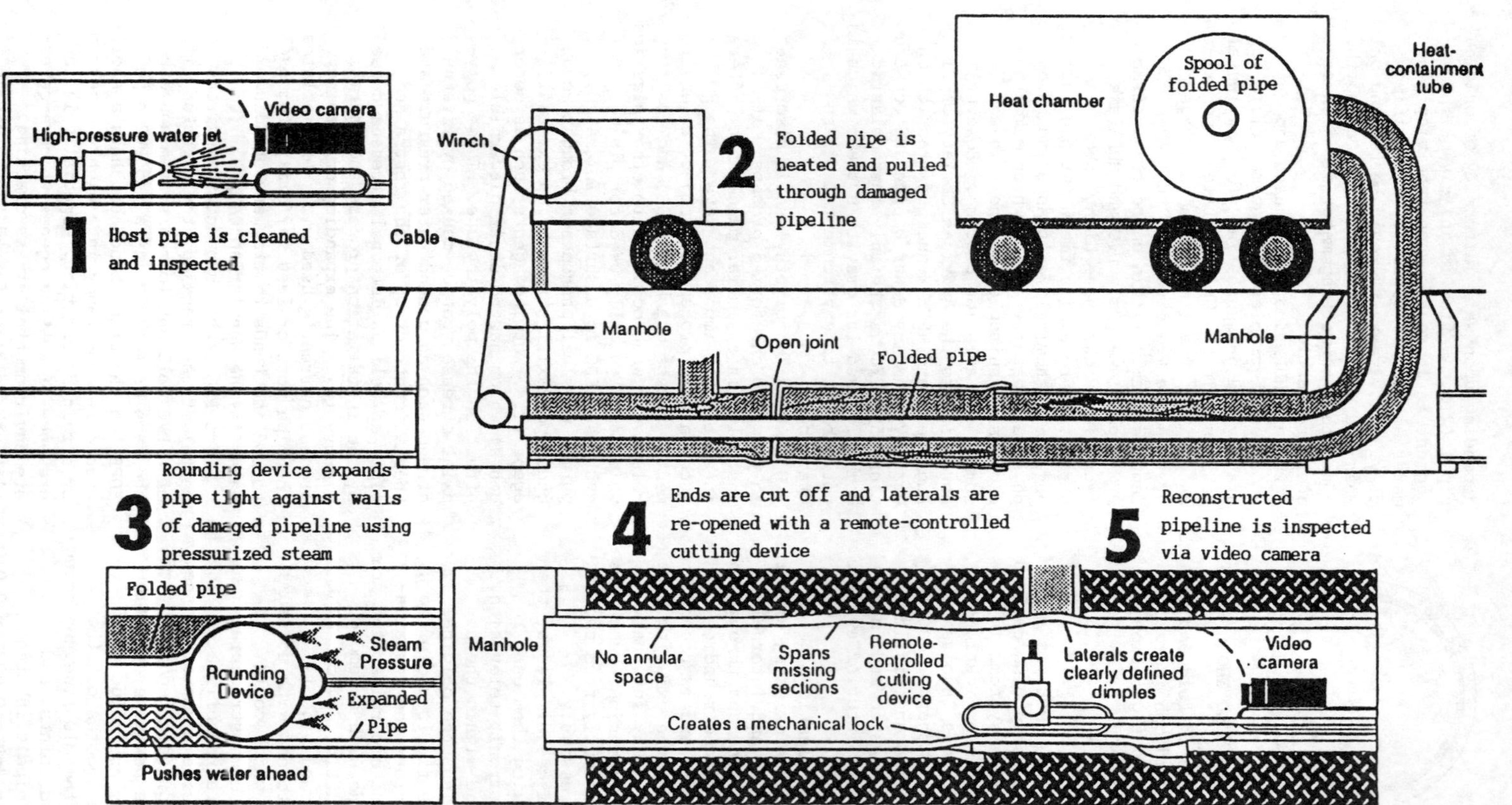

Figure 6. How PVC Fold and formed is installed.

currently being used in the United States, it does not yet have a time tested, proven track record.

4 SLIPLINING

Standard sliplining techniques are simply the pushing or pulling of a smaller diameter (usually polyethylene) pipe inside of an existing deteriorated line. Although the general nature of this process is sometimes referred to as trenchless, excavation may be required for insertion pits and for reinstating lateral side connections. When a system is sliplined, there is an annular space between the new pipe and the old host pipe which should be filled. When filled, a low strength Portland cement grout is generally used. Grouting greatly increases the in-place structural strength and extends the service life. This process leaves a smaller diameter pipe as the finished product which may decrease flow over the original pipe. Standard sliplining is a widely used pipe rehabilitation method.

Segmental sliplining is a technique using relatively short joints of pre-manufactured pipe and inserting them into an existing deteriorated line. This has the usual disadvantages of sliplining, including some excavation, plus the added unknown of multiple joints. Like standard sliplining, the annular space should be filled. Lateral reinstatement in small diameter pipe must be done externally and requires additional excavation. Flow is almost always reduced, and insertion can cause high trauma. An advantage of segmental sliplining is that in many cases a deteriorated line can be reconstructed without having to bypass the existing flow. This is especially beneficial for larger diameter pipe because bypassing large quantities can be difficult. As with traditional sliplining, only straight lines or pipes with gentle bends can be reconstructed.

Another sliplining technique, in the developmental stage, uses polyethylene and reduces the diameter by swaging or drawdown and then, while at the reduced diameter, it is inserted into an existing line. After the insertion, it is expanded against the walls of the host pipe. This process seems to be more applicable to gas lines although it may be suitable for certain sewer collection systems. Under ideal conditions, this process can produce a new polyethylene pipe inside a host pipe with no or little annular space. Draw-

backs associated with this process can be failure to initially reduce the diameter enough for complete insertion, inability to expand directly against the walls of the host pipe, and excavations which are required for an insertion pit, receiving pit, and for lateral side connections. Some success with internal reinstatement of side connections has been reported from the field, but so far, external reconnection seems the norm.

The primary advantage of the swaging or drawdown process is that it eliminates one of the big disadvantages of direct sliplining, namely decreased flow. If the HDPE can be installed correctly and is expanded against the walls of the old pipe, hydraulic capacity can be restored or increased over the pre-rehabilitated condition.

Spiral winding is a variation of sliplining which involves the installation of ribbed profile PVC through a manhole utilizing a special spiral winding machine which rolls up to 8" strips into the pipe to be rehabilitated. The strips are either glued or mechanically locked into place. Grouting of the annular space is required as with standard sliplining; however, because of the numerous joints and low pipe stiffness, grout pumping pressures must be low. Excavation is required for reinstatement of service connections in non-man entry sizes.

Pipe bursting is another form of sliplining. The preparation for pipe bursting is generally the same as for standard sliplining, and excavation for access is usually required. A bursting device is pushed and/or pulled through the existing line and through either impact, hydraulic expanding, or rotary drilling, the existing pipe is shattered and pushed into the surrounding soil. As the bursting head progresses through, it pulls an attached HDPE pipe in place. Pipe bursting replacement rates vary widely depending on the soil conditions, process method, and structural stability of the existing pipe. The only types of pipe that can be rehabilitated using this method are brittle, nonreinforced materials such as nonreinforced concrete or vitrified clay. The chance of environmental stress cracking occurring may be increased because of the inherent trauma inflicted upon the new HDPE pipe during insertion. This can be overcome by sliplining the installed pipe with another plastic pipe. Since the bursting process destroys any side connections, excavation is required for each lateral. Vibration can be a potential for damage to other

utilities located nearby, and if the
bursting head becomes stuck within the
pipe under repair, an excavation is
required for a point repair.

This process can, in certain conditions,
actually install a new pipe that is
slightly larger than the original, which
increases flow capacity.

Of the four types of reconstruction,
only two offer pipe replacement without
excavation, cured-in-place pipe and fold
and formed pipe. Some excavation is
required for sliplining, although it could
be considered semi-trenchless construction. The need to reconstruct and extend
the life of our in-place pipeline infrastructure continues to grow. The problems
associated with pipeline rehabilitation
are complex and interrelated with no two
projects being exactly alike. The choice
of technologies must be studied carefully,
for each specific application, in order to
provide a cost effective solution for
renewing our existing piping systems.

REFERENCES

(1) Utah State University, 1988.
        Contribution of Insitupipe to the
        structural integrity of broken
        rigid buried pipes. Logan, Utah.
(2) United States Environmental
        Protection Agency, 1983.
        Demonstration of sewer relining by
        the Insituform process, Northbrook,
        Illinois. Cincinnati, Ohio.
(3) Sverdrup Corporation and Southeast
        Environmental Services. 1990.
        Sewer segment flow monitoring study
        to establish Manning roughness
        coefficients. St. Louis, Missouri
        and Murfreesboro, Tennessee.
(4) Utah State University, 1975.
        Hydraulic characteristics of PVC
        sewer pipe in sanitary sewers.
        Logan, Utah.
(5) ASCE Manuals and Reports on
        Engineering Practice - No. 60, WPCF
        Manual of Practice - No. FD-5
        Gravity sanitary sewer design and
        construction by the American
        Society of Civil Engineers and the
        Water Pollution Control Federation
(6) Howard, J. B. 1964. Stress
        cracking. In E. Byer (ed.)
        Engineering design for plastics,
        Chapter 11. New York: Reinhold

Structural Performance of Flexible Pipes, Sargand, Mitchell & Hurd (eds) © 1990 Balkema, Rotterdam. ISBN 90 6191 165 6

Soil-pipeline interaction in a pipeline with prescribed displacements

A.P.S.Selvadurai & M.C.Au
Department of Civil Engineering, Carleton University, Ottawa, Ont., Canada

S.B.Shinde
Geotechnical Section, Esso Resources, Calgary, Alb., Canada

ABSTRACT: The present paper develops a theoretical model for the study of soil-pipeline interaction in a long distance pipeline with a soft backfill. The interaction is induced by a random variation in pipeline displacements. The soil model that is utilized in the analysis is a two-parameter Vlazov-type or Pasternak-type foundation where there is shear continuity between the one-dimensional Winkler-type elements. In the specific example treated in the paper, the interaction between the soil and the pipeline is induced by prescribing a displacement which varies in a random fashion. The paper presents numerical results which illustrate the flexural moment distribution at salient locations along the pipeline.

1. INTRODUCTION

Buried pipelines are used for the offshore and on-shore transportation of oil, natural gas, coal slurries, mine tailings and water. Such pipelines are built and operated according to standard codes to ensure safe and economical operation for substantial periods of time (Ariman et al. 1979; Pickell, 1983; Jayapalan, 1985). Although the methodologies used for designing such long distance pipelines have been supported by extensive research work over the years, opportunities do exist to further refine existing techniques so that cost effective and reliable designs can be achieved. In this regard, the analysis of soil-pipeline interaction is an important aspect in the study of buried long distance pipelines (Selvadurai et al. 1983; Selvadurai 1983, 1984, 1985, 1988; Selvadurai and Pang, 1988). In conventional treatments of a soil-pipeline interaction related to a buried pipeline, the response of the soil is idealized in terms of one-dimensional elements which possess linear or non-linear force-deformation relationships. Other higher order models such as multi-parameter soil models incorporate the load transfer mechanisms between the one-dimensional elements. In advanced treatments of the soil-pipeline interaction, the soil is

represented by elastic and elastic-plastic continuum models and takes into consideration the influence of bonding, separation and slip at the soil-pipeline interface. The simple one-dimensional Winkler-type soil models are not exact representations of the constitutive behaviour of soils. The parameters describing such models are system parameters which depend on the particular soil-pipeline system. Other limitations of the model include the inability to accommodate adequately the transfer of external loads to the pipeline and precisely defining the support condition for a pipeline with a three dimensional configuration. These models are however relatively easy to incorporate into numerical codes for soil-pipeline interaction. The continuum models offer more adequate representations of the soil mass surrounding a pipeline. Such models can be incorporated in numerical codes which utilize finite element and boundary element techniques, which accommodate linear, nonlinear and time-dependent phenomena. The constitutive properties which characterize continuum soil models are not system parameters and can be obtained separately from conventional geotechnical tests. The analysis of soil-pipeline interaction by utilizing continuum models of soil behaviour is certainly non-routine and requires extensive computational effort. Con-

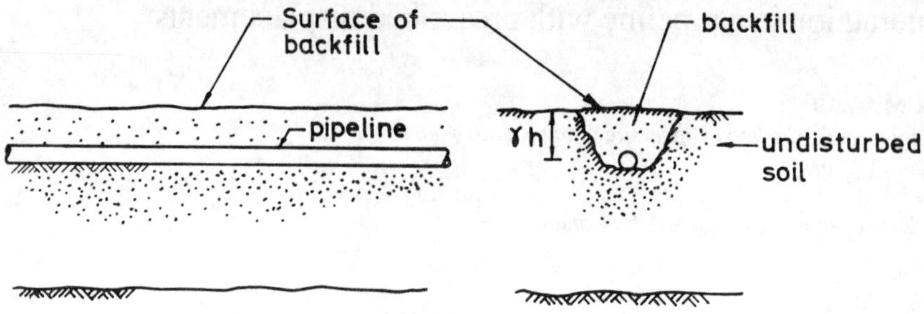

Figure 1. Buried pipeline with soft backfill

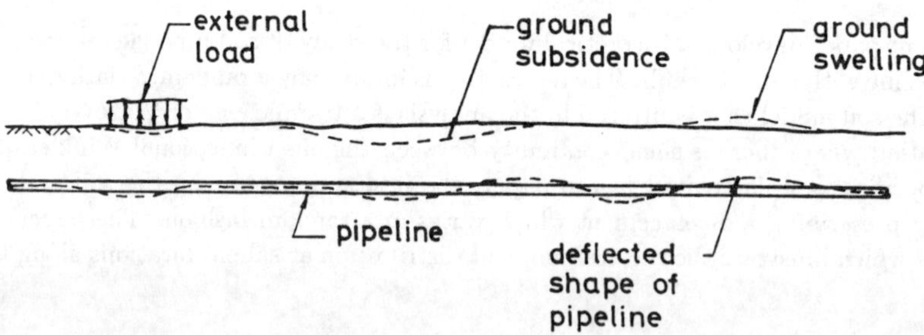

Figure 2. Soil-pipeline interaction due to ground movement

sequently it is desirable to investigate an intermediate class of soil models which have the computational flexibility of the one dimensional models but are capable of representing the mechanical behaviour of soils in a manner similar to that provided by the continuum models. The simplest form of such an intermediate soil model which can be used in a soil-pipeline interaction problem is the two-parameter or Pasternak-type soil model which achieves the continuity in the soil response by introducing a shear interaction layer between the one-dimensional elements (see e.g., Kerr, 1964; Selvadurai, 1979). The response of this idealized soil model can also be interpreted by appeal to the mechanics of a soil layer of finite thickness which deforms in a constrained fashion (Vlazov and Leontiev, 1966).

The present paper deals with the analysis of flexure of a pipeline which rests in contact with elastic soil layer which is idealized by a two parameter continuous elastic medium. The pipe rests on the surface of such a medium and achieves contact with the medium over a portion of its circular boundary. The contact is achieved by a

backfill region which is considered to be a soft material and which only provides a surcharge load and does not possess any stiffness characteristics (Figure 1). This particular situation could be encountered in a pipelining operation where the trench backfill is loosely placed without adequate compaction. The linear elastostatic finite element technique is used to examine the flexural interaction between a long-distance pipeline and the surrounding soil. The interaction process is induced by non-uniform pipe movements in the vertical direction, which may be induced by external loads, ground subsidence and ground swelling (Figure 2). To simulate such actions, the pipeline displacements are prescribed in a pseudo-random fashion. The Galerkin finite element scheme is used to develop results for the distribution of flexural moments along the length of the pipeline. The numerical results presented in the paper illustrates the manner in which these flexural moments are influenced by the relative stiffness of the soil-pipeline system and the degree of shear transfer between the supporting soil elements.

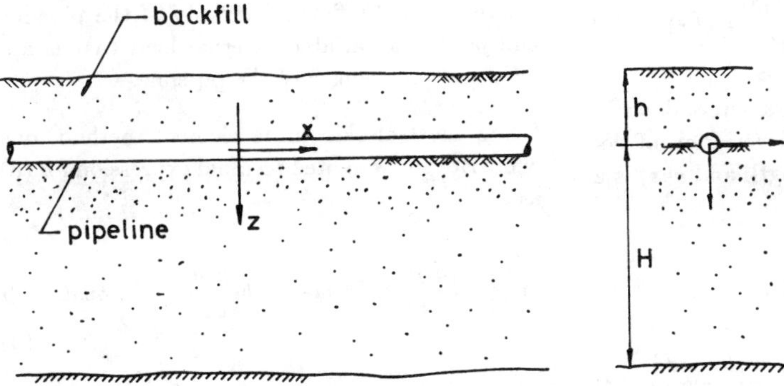

Figure 3. Geometry of the buried pipeline problem

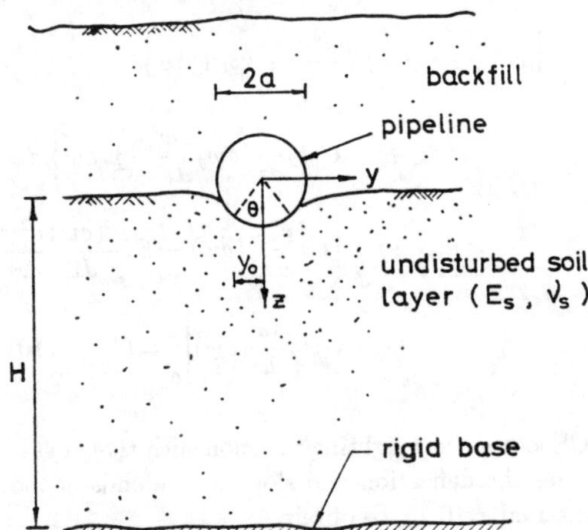

Figure 4. Cross-sectional view of the pipeline; influence of three-dimensional geometry of the pipeline

2. ANALYSIS

We consider the problem of a pipeline of infinite length which rests on an idealized two-parameter elastic foundation. When the Vlazov and Leontiev (1966) idealization of the two-parameter foundation is invoked the modelling can be related to the problem of a circular pipeline which rests on a constrained soil layer of thickness H. The pipeline is maintained in contact with the soil layer by the loosely placed backfill which does not participate in the interaction process except through the provision of a surcharge effect (Figure 3). The contact region between the pipeline and the underlying soil layer is a variable pa-

rameter along the length of the pipeline. The extent of the contact is also influenced by the soil-pipeline interaction process. Consequently in the complete analysis of the soil-pipeline interaction problem related to a constrained soil layer or a two parameter foundation, the contact region should be determined as a part of the solution. This entails a considerable degree of analytical and numerical work which can only be justified in situations where the pipeline is subjected to loading influences which are fairly well defined (e.g., single external loads, pipe movement at a prescribed location, etc.). When dealing with completely random pipe displacement patterns it is justifiable to assume the contact region to be constant and to ascribe plausible values for this parameter. An alternative to this approach would be to assign the contact region to be a variable parameter along the length of the pipeline; this is relegated to future investigation. It may be noted that the self weight of the pipeline or the weight of a surcharge on the pipeline will induce uniform indentation along the pipeline which will establish the contact region. The variations in the value of contact region is caused by the ground movement induced interaction between the pipeline and the soil.

Consider the cross section of the pipeline which is in contact with the two parameter solid (Figure 4). The resultant of contact pressures and concentrated edge reactions, acting per unit length along the length pipe (in the x-direction) is denoted by $R(x)$. Considering the analysis of this problem (Selvadurai and Au, 1991) it can be shown that

137

$$R(x) = kb_1 w(x) - G_p b_2 \frac{d^2 w(x)}{dx^2} + q(x) \qquad (1)$$

where k and G_p are the two parameters describing the soil model; $w(x)$ is the variation of deflection of the pipe along its length and $q(x)$ is a reactive force component given by

$$q(x) \;=\; -2\lambda G_p a \left(1 - \cos\theta\right)$$

$$-\; 2ka^2 \left\{ \sin\theta - \frac{\theta}{2} - \frac{\sin\theta\cos\theta}{2} \right\} \qquad (2)$$

where 2θ is the contact angle and $\lambda = \sqrt{k/G_p}$. Also in (1) b_i $(i = 1,2)$ are equivalent width parameters defined by

$$b_1 = 2\left(y_0 + \frac{1}{\lambda}\right) \qquad (3)$$

$$b_2 = 2y_0 \qquad (4)$$

$$y_0 = a\sin\theta \qquad (5)$$

Also for a Vlazov-type foundation layer which has constrained displacements in the horizontal direction it can be shown that

$$k = \frac{E_s}{H\left(1 + \nu_s\right)\left(1 - 2\nu_s\right)} \qquad (6)$$

$$G_p = \frac{E_s H}{6\left(1 + \nu_s\right)} \qquad (7)$$

where E_s and ν_s are Young's modulus and Poisson's ratio respectively and H is the thickness of the soil layer.

Consider the flexure of a circular pipeline which rests in contact with a constrained elastic layer of the Vlazov- or Pasternak-type. The contact region between the pipe and the elastic layer is assumed to be defined by a constant value of θ. The complete interaction equation for the pipeline resting on the constrained elastic layer can be written as

$$EI\frac{d^4 w}{dx^4} + kb_1 w - G_p b_2 \frac{d^2 w}{dx^2} = p(x) - q(x) = p^* \qquad (8)$$

where EI is the flexural rigidity of the pipeline and $p(x)$ is a variable external load that is applied per unit length of the pipeline.

Using the Galerkin finite element method, one can write the weighted residual expression for (8) as

$$\int_0^\ell \left\{ EI\frac{d^4 w}{dx^4} + kb_1 w - G_p b_2 \frac{d^2 w}{dx^2} - p^* \right\} \delta w\, dx = 0 \qquad (9)$$

where δw is the weighting function and ℓ is the length of the pipeline under consideration. Integrating by parts, twice, the first and third terms in (9) we obtain

$$\int_0^\ell \left\{ EI\frac{d^2}{dx^2}\,(\delta w)\;\frac{d^2 w}{dx^2} + kb_1 w(\delta w) \right.$$

$$+\quad G_p b_2 \frac{d}{dx}(\delta w)\frac{dw}{dx} - p^*\delta w \Bigg\} dx$$

$$+\quad \left[EI\frac{d^3 w}{dx^3}(\delta w) - EI\frac{d(\delta w)}{dx}\frac{d^2 w}{dx^2} \right.$$

$$-\quad G_p b_2 \frac{dw}{dx}(\delta w) \Bigg]_0^\ell = 0 \qquad (10)$$

Choosing the weighting function such that it satisfies the deflection and slope at the ends of the interval $x\epsilon(0,\ell)$ we obtain

$$\int_0^\ell \left[EI\frac{d^2}{dx^2}\,(\delta w)\;\frac{d^2 w}{dx^2} + G_p b_2 \frac{d}{dx}(\delta w)\frac{dw}{dx} \right.$$

$$+\quad kb_1(\delta w)w - p^*(\delta w)\big]\, dx = 0 \qquad (11)$$

Alternatively (11) can be written as

$$\delta\pi = 0 \qquad (12)$$

where

$$\pi \;=\; \frac{1}{2}\int_0^\ell \left\{ EI\left(\frac{d^2 w}{dx^2}\right)^2 + G_p b_2 \left(\frac{dw}{dx}\right)^2 \right.$$

$$+\quad kb_1 w^2 \Big\} dx - \int_0^\ell p^* w\, dx \qquad (13)$$

which is the required expression for the finite element formulation. The result (13) is also the total potential energy function for the pipeline-constrained elastic layer system. The finite element formulation can be developed either by using (11) or (13). The details of the finite element formulation will be presented elsewhere (Selvadurai and Au, 1991).

3. A SOIL-PIPELINE INTERACTION PROBLEM

As an example of the application of the above developments we consider the problem of a long-distance buried pipeline in a ground movement zone. We assume that the ground displacements at prescribed locations occur in a random fashion. This particular type of problem is important to the assessment of the state of stress in pipelines located in a ground subsidence or ground swelling zone. Actual monitoring of pipeline movements can be carried out either directly or via devices which traverse the length of the pipeline. In such situations the ground movements are the only agencies responsible for the soil-pipeline interaction.

From the theory of statistics, the simulation of actual experimental data is achieved by a set of random numbers which can be generated. In practice two other methods are commonly used. One involves using lists of tabulated random numbers. An alternative is to generate a sequence of apparently random numbers by using an algebraic formulae. The use of such a technique seems to contradict the requirement of a random sequence; consequently the generated sequence is referred to as a "pseudo-random number sequence". The most commonly used procedure takes the form (Downham and Roberts, 1967; Davies and Goldsmith, 1972)

$$R_{i+1} = AR_i + B \, mod(M) \; ; \quad i = 0, 1, 2, \cdots \quad (14)$$

where R_i, A, B and M are integers with M being the highest integer. This yields a sequence of numbers which fall evenly in the range 0 to M. Using this procedure a cycle of numbers can be generated with a very long period.

In the context of the soil-pipeline interaction prob-lem, we employ the random number as the magnitude of the pipe deflection along the pipeline. The pipeline has a diameter of 9 ins. and a pipe wall thickness of 0.25 ins. For the analysis of the problem, we set the semi contact angle at 45°. The pipeline material has a Young's modulus of 30×10^6 lb/in^2, such that the flexural rigidity of the pipeline $EI = 2.15 \times 10^9$ lb/in^2. The pipe is in contact with a constrained soil layer of thickness 100 in. The pipeline is modelled by a series of 100 in. long beam elements. In order to simulate a reasonably long pipeline, 900 such pipe elements are used. This is equivalent to a pipeline of length 1.4 miles. The Poisson's ratio for the soil layer is set equal to 0.25. The interaction between the pipeline and the soil is influenced by the relative stiffness of the pipeline-soil system. This is characterized by the non-dimensional relative stiffness parameters R given by

$$R = \frac{EI}{4E_s a^4} \qquad (15)$$

The maximum and minimum deflections induced in the pipeline by random ground movement are 1.0 in. and -1.0 in. respectively. The Galerkin finite element technique described in the previous section is applied to evaluate the variation in the bending moments along the pipeline due the soil-pipeline interaction. The Figures 5 to 8 illustrate the manner in which the flexural movements along the pipeline are influenced by the relative stiffness of the pipe-soil system.

4. CONCLUSIONS

The study of soil-pipeline interaction induced by ground movements is important to the efficient design of a pipeline. The accurate modelling of such interaction processes can also be useful in identifying the in-service state of stress in a pipeline. A number of methodologies are available for representing the behaviour of the soil for purposes of soil-pipeline interaction. The very elementary models of the Winkler type have been extensively used in the study of soil-pipeline interaction. The more sophisticated continuum models offer techniques for accurately representing the soil-pipeline interaction process. An intermediate class of soil models accounts for the continuous behaviour of the soil support by ei-

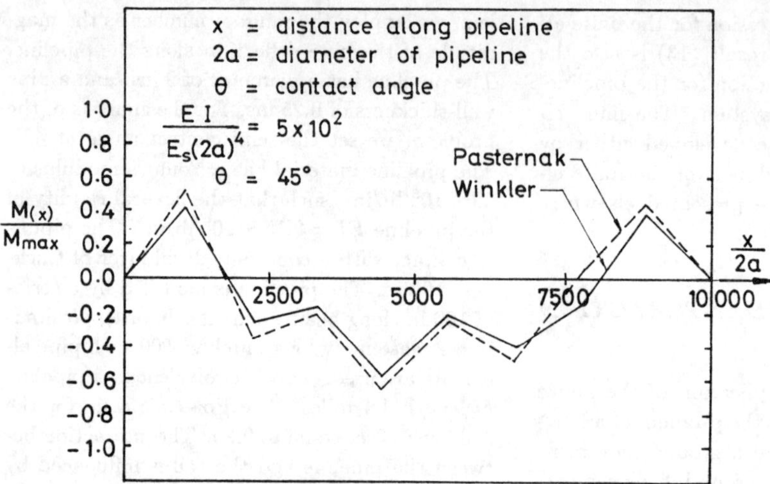

Figure 5. Distribution of flexural moments at selected locations of the pipeline subjected to ground movement.

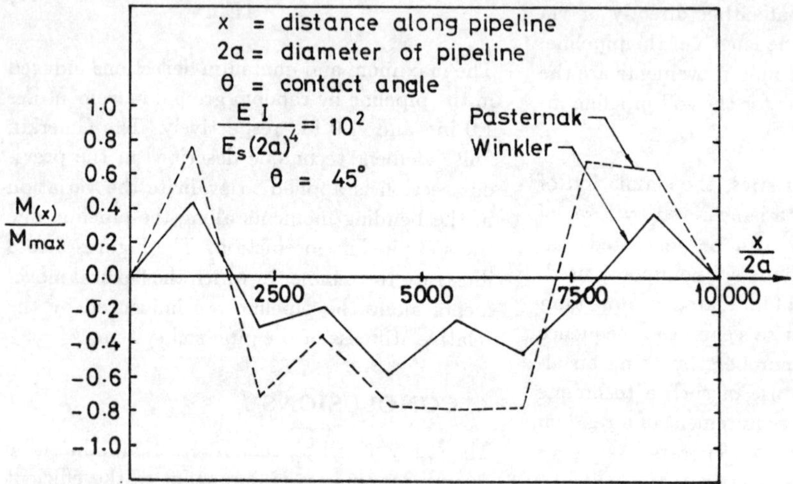

Figure 6. Distribution of flexural moments at selected locations of the pipeline subjected to ground movement.

ther constraining the behaviour of the continuum or incorporating shear interaction capability between the individual Winkler elements. This paper examines the flexural interaction of a pipeline resting on such an elastic medium. The analysis accounts for, in an approximate fashion, the influence of three dimensional effects in such a soil-pipeline interaction problem. The analysis of the flexural interaction process yields a basic dif-ferential equation which is solved via a Galerkin finite element scheme. The finite element scheme is applied to examine the flexural moments developed in a pipeline which experiences random displacements due to ground movements. The numerical results presented in the paper illustrates the manner in which the flexural moments in the pipeline are influenced by the continuum behaviour of the supporting soil. The numeri-

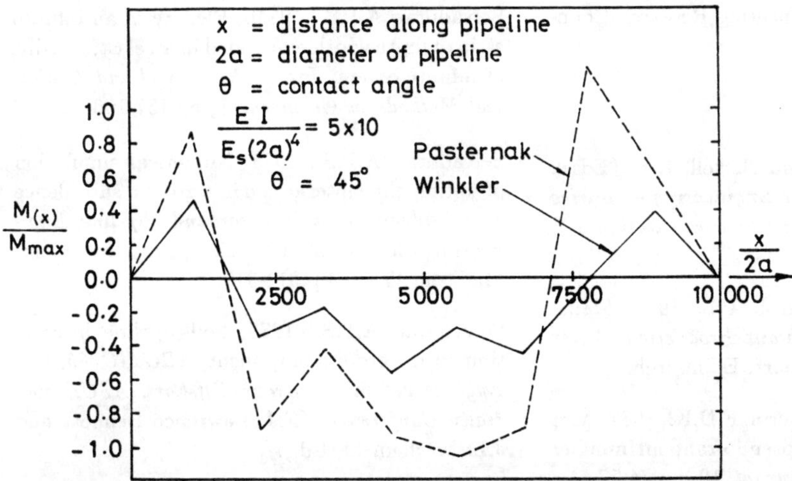

Figure 7. Distribution of flexural moments at selected locations of the pipeline subjected to ground movement.

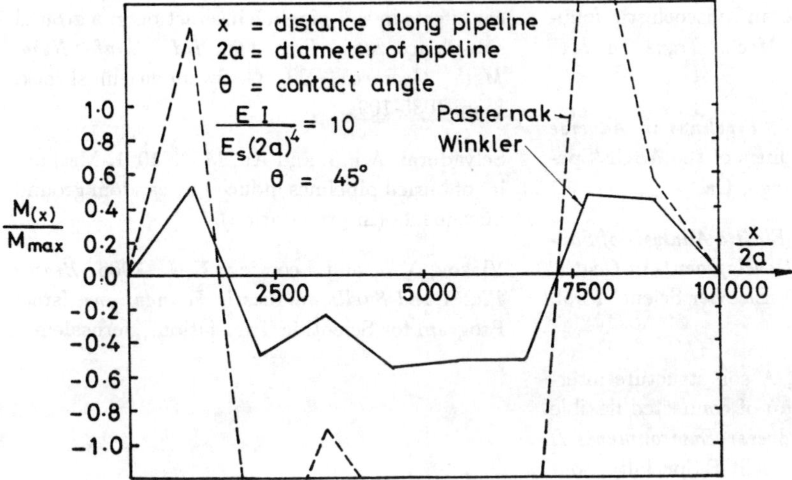

Figure 8. Distribution of flexural moments at selected locations of the pipeline subjected to ground movement.

cal results offer a comparison between the flexural moment distributions derived by the Winkler model and the continuous soil model. It is evident that as the relative stiffness of the pipe-soil system increases, the two sets of results give reasonably close values. As the relative stiffness of the pipe-soil system decreases, the two soil support models give widely varying distributions for the flexural moments. The results of this paper suggests that the interpretation of the state of stress in a pipeline as inferred via a set of measured pipe displacements needs to be approached with caution. Such interpretations invariably require an accurate modelling of the deformational behaviour of the soil.

ACKNOWLEDGEMENTS

The work described in this paper was supported by a Research Grant A3866 awarded by the National Sciences and Engineering Research Council of Canada.

REFERENCES

Ariman, T., Liu, S.C. and Nickell, R.E. (Eds.), 1979. *Lifeline Earthquake Engineering — Buried Pipelines, Seismic Risk and Instrumentation*, PVP 34 Amer. Soc. Mech. Engrs.

Davies, O.L. and Goldsmith, P.L., 1972. *Statistical Methods in Research and Production*, Oliver and Boyd, Tweeddale Court, Edinburgh.

Downham, D.Y. and Roberts, F.D.K., 1967. Multiplicative congruential pseudo-random number generators, *Computer Journal*, 10, pp.74-77.

Jeyapalan, J.K. (Ed.), 1985, *Advances in Underground Pipeline Engineering*, Proc. Int. Conf., Madison, Wisc.

Kerr, A.D., 1964, Elastic and viscoelastic foundation models, *J. Appl. Mech. Trans. ASME*, 31, pp.491-498.

Pickell, M.B. (Ed.), 1983, *Pipelines in Adverse Environments II*, Proceedings of the ASCE Specialty Conference, San Diego, Ca.

Selvadurai, A.P.S., 1979, *Elastic Analysis of Soil-Foundation Interaction, Developments in Geotechnical Engineering, Vol.17*, Elsevier Scient. Publ. Co., Amsterdam.

Selvadurai, A.P.S., 1983, A soil structure interaction problem for a group of connected flexible pipelines, *Pipelines in Adverse Environments II* (Ed. Pickell, M.B.), Proc. ASCE Specialty Conference, San Diego, Ca., pp.199-211.

Selvadurai, A.P.S., Lee, J.J., Todesschimi, R.A.A. and Somes, N.F., 1983, Lateral soil resistance in soil-pipeline interaction, *Pipelines in Adverse Environments II*, Proc. ASCE Specialty Conference, San Diego, Ca., pp.259-278.

Selvadurai, A.P.S., 1984a, Elastostatic soil-structure interaction problems for embedded flexible and rigid structures, *Proc. CSCE Specialty Conference on Computer Methods in Offshort Engineering*, Halifax, N.S., pp.321-350.

Selvadurai, A.P.S., 1984b, Flexure of an infinite strip of finite width embedded in an elastic medium of infinite extent, *Int. J. Numerical and Analytical Methods in Geomech.*, 8, pp.157-166.

Selvadurai, A.P.S., 1985, Numerical simulation of soil-pipeline interaction in a ground subsidence zone, *Advances in Underground Pipeline Engineering*, (Jeyapalan, J.K., Ed.), Proc. Int. Conf. Madison, Wisc., pp.311-319.

Selvadurai, A.P.S., 1985, Soil-pipeline interaction during ground movement, *ARCTIC '85, Civil Engineering in the Arctic Offshort, ASCE Specialty Conference*, (F.J. Lawrence Bennett and J.L. Machemehl, Ed.).

Selvadurai, A.P.S., 1988, Mechanics of soil-pipeline interaction, *Proc. CSCE Annual Conference*, Calgary, Alberta, 3, pp.151-173.

Selvadurai, A.P.S. and Pang, S., 1988, Nonlinear effects in soil-pipeline interaction in a ground subsidence zone, *Proc. 6th Int. Conf. Num. Meth. Geomech*, (Ed. G. Swoboda) Innsbruck, 3, pp.1085-1094.

Selvadurai, A.P.S. and Au, M.C., 1991, Mechanics of buried pipelines induced by random ground movements (in preparation).

Vlazov, V.Z. and Leontiev, U.N., 1966, *Beams Plates and Shells on Elastic Foundations*, Israel Program for Scientific Translations, Jerusalem.

Structural Performance of Flexible Pipes, Sargand, Mitchell & Hurd (eds) © 1990 Balkema, Rotterdam. ISBN 90 6191 165 6

Disposal of stormwater runoff using recharge systems

Corwin L.Tracy
National Corrugated Steel Pipe Association, Washington, D.C., USA

ABSTRACT: The increased development of the urban areas in the past decade of all of the major metropolitan areas in the United States has prompted the need for innovative methods of dealing with stormwater runoff. Also, the advent of the zero increase in stormwater runoff criteria used by many of our governmental agencies has resulted in engineers using a number of different and unique designs for retention, detention and recharge systems facilities to meet this criteria. The intent of this paper is to provide current state-of-the-art information dealing with facilities constructed for subsurface disposal of stormwater and specifically covering recharge/infiltration systems. Four types of these infiltration/recharge systems are: 1. Spreading basins; 2. Wells; 3. Trenches and 4. Combination underground detention chamber and recharge facility. A number of different types of material and construction specifications used by municipalities, counties and drainage control districts located in different areas of the United States are discussed. The many economical, ecological and environmental advantages realized when recharge systems are used for stormwater runoff disposal are pointed out, plus a few of the disadvantages.

1 INTRODUCTION

In the past year or so, our nation's population, ecological and environmental awareness, has increased many times over. With the passing anniversary of Earth Day on April 22, 1990, focus was on how we are tragically polluting our environment. One of the areas, I believe, of utmost concern is the polluting of our potable water. For the past two or three years, we have evidenced droughts in several areas of our country. This event has commanded attention, especially the attention of those people residing in those areas. These people, plus many others in this country, realize the time has come to act if we want our families and future generations to survive.

In the words of Richard F. Curtis, Chairman and CEO, Environ Technologies Corporation, "...When we were water-wealthy, we could afford to regard water as an inalienable birthright...we can no longer be so cavalier. Remember, water and life are inseparable...water is not destroyed, only used. We cannot destroy it, but we can make it unusable and that amounts to the same thing. Water is something we take for granted, like breathing in and out or watching the sun come up. But we cannot afford to take it for granted much longer. For many years, knowingly or unknowingly, we have been deferring the costs of our water use practices into the future. Now the future has arrived and the bill is ... about to come due!" (1).

Past stormwater management policies placed emphasis on quick removal of stormwater through "positive drainage"--the disposal of water through a closed system into an open stream or body of water. While these positive systems are highly efficient in removing quantities of water, they do not allow for the removal of stormwater pollutants before entering a surface water quality. (2).

In addition to pollution, these "positive systems" create other problems such as high peak flows; excessive erosion; increased flooding and lowering of the water table. Nature, through a system of bogs, swamps, forested areas and undulating terrain intended that the water soak back into the earth. One approach which would help emulate nature's practices is to direct stormwater back into the soil. (3).

Recent years have seen artifical recharge

143

of ground water become one of the high priority items for federal, state and local funding in arid parts of the United States as well as in many arid countries. This has been due to problems created by increasing development of irrigation and public water supply from ground water and consequent lowering of water tables to uneconomically low levels and by degradation of ground water quality. The use of surplus surface water and treated waste water has gained high popularity in recent years to recharge the ground water basins and is expected to increase on an exponential scale in the future as the industralized nations grow and developing countries modernize and require more water per capita. (4).

With this current emerging water crisis, we are going to focus our comments in this paper on management of stormwater runoff through the use of recharge or infiltration systems and the advantages to be gained by use of same.

Advantages to be gained through the use of a recharge/infiltration system are:

1   conservation--ground water replenishment;
2   reduced need for treatment of stormwater;
3   reduced or zero increase in discharge;
4   reduced subsidence due to ground water withdrawal;
5   provides sea water barriers in coastal aquifers;
6   reduced need for pumping stations or extensive outlet piping or drainage channels.

Resultant disadvantages: (5) The disadvantages which may result through the use of a recharge/infiltration system are:

1   pollution of ground water table;
2   with some types, unusual high maintenance;
3   frost penetration;
4   site suitability--design problem;
5   public acceptance/suitability.

Types of Recharge/Infiltration systems:
There are four types of recharge/infiltration systems commonly used today. They are:

1   spreading basins;
2   wells;
3   trenches;
4   combination underground detention chamber and recharge facility.

2   SPREADING BASINS

Water is released from the source to the surface of a basin where it percolates through the ground into the aquifer. The spreading basins are usually in or near a streambed and are formed by creating dikes using the material from the bottom of the basin.

When ample space is available and other criteria are satisfied, the use of spreading basins can provide a relatively inexpensive solution to stormwater disposal in terms of cost per unit volume of water drained. Space is often available within areas of the right-of-way, such as highway interchanges, or on non-used portions of residential or commercial developments. A typical spreading basin is shown in Figure 1. (3).

In some communities, infiltration basins have been integrated with attractive parks and/or recreation areas. This dual role of the basin benefits both the facility and the public.

Spreading basin surfaces are sometimes scarified to break up silt deposits and restore topsoil porosity. This should be accomplished after all sediment has been removed from the basin floor. However, this operation can be eliminated by the establishment of grass cover on the basin floor and slopes. Such cover helps maintain soil porosity.

Algae or bacterial growth can inhibit infiltration. While chlorination of the runoff water can solve this problem, it is more practical to make certain that the basin is permitted to dry out between storms and during summer months.

Cleanout frequency of spreading basins will depend on whether they are vegetated or non-vegetated and will be a function of their storage capacity, infiltration charactreistics, volume of inflow and sediment load. Holding ponds or sedimentation basins can be used to reduce maintenance in conjunction with spreading basins by settling out suspended solids before the water is released into the spreading basins.

Among the negative aspects of spreading basins are their susceptibility to early clogging and sedimentation and the considerable surface land areas required for their construction. This method requires the most land use of any of the recharge methods and may be a significant factor if the cost of acquiring the land is too high. Basins also present a security problem due to exposed standing water and a potential for insect breeding. (6).

3   WELLS

Wells have been used for many decades for conducting water into the ground. We

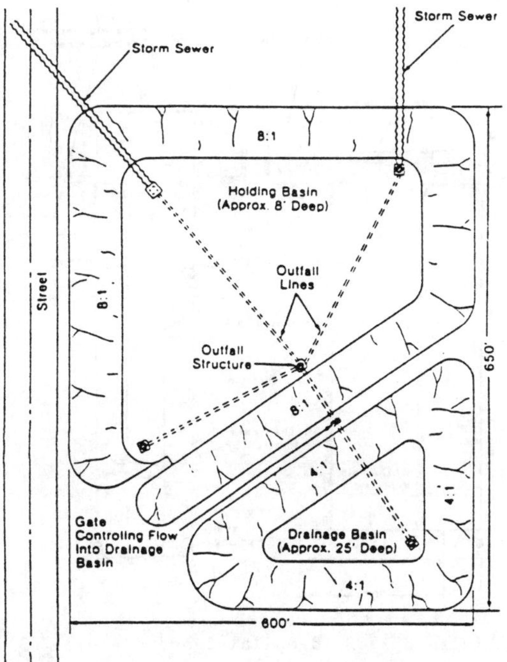

Figure 1 - spreading basin (3)

have divided them into two categories-
dry and recharge. The oldest is perhaps
the dry well which is a small-diameter
hole or pit dug into the ground for the
disposal of water that has no natural
drainage. A dry well is usually filled
with pea gravel, coarse sand, or other aggregate, or contains a slotted or perforated pipe, backfilled with materials which
allow water to penetrate and soak into the
ground, while preventing collapse of the
walls. (6).

Wells require a minimum of space and may
be designed with very little unsightly surface structure. They can be extended
through impervious soils down to permeable
sand or gravel and will drain a small area
fairly rapidly when surface runoff is of
satisfactory quality.

Examples of typical dry wells are shown
in Figures 2, 3 and 4.

The use of recharge wells in the coastal
areas of Florida for stormwater disposal has
increased substantially over the past few
years. This has happened since the installation of positive systems has been virtu-
ally discontinued and also this technique
is used to prevent or slow salt water intru-
sion. This type well is an open-ended vertical pipe which discharges stormwater to
that portion of the aquifer containing a
minimum of 1500 ppm of chlorine (consider-

ed salt water). The installation of this
structure requires a permit from the State
of Florida Department of Environmental
Regulation. Such a permit is issued only
after the state receives comments from the
Dade County Department of Environmental
Resources Management (D.E.R.M.). (2).

An example of this type recharge well and
the grease interceptor facility adjoining
it is shown in Figure 5.

The same clogging and chemical reactions
that retard basin and trench infiltration
will affect wells to an even greater extent. One problem unique to wells is
chemical encrustation of the casing, with
consequent blocking of the perforations or
slots in the wall casing. Alternative wetting and drying builds up a scale of water
soluble minerals which can be broken up or
dissolved by jetting, acid treatments, or
other procedures.

Wells may pollute ground water supplies,
therefore, local health agencies should be
contacted before proceeding with such installations. The capacity for drainage is
difficult to predict with wells in a single
area having different rates of infiltration.

For these reasons, wells should generally
be considered for disposal of small quani-
ties of water or as a supplement to recharge basins or some other type of disposal system. (3).

4 TRENCHES

The infiltration trench is a modification
of the spreading basin. The addition of
perforated pipe to the spreading basin concept increases the exfiltration from the
trench by more than 100 times that of conventional "French drains" or dry wells
which are limited by cross-sectional area.
It also serves the function of collecting
sediment before it can enter the coarse
rock backfill. As collected, sediments
are distributed throughout the length of
the free-flow area and clogging is minimized. For example, the sediment-laden
water must flow through the cross-section
of the conventional French drain to flood
the trench and gain access to the trench
wall. The perforated pipe distributes the
water immediately for its full length, providing immediate access to the trench wall.

For perforated corrugated metal pipes,
3/8-inch diameter perforations uniformly
spaced around the full periphery of a pipe
are desirable. Not less than 30 perforations per square foot of pipe surface
should be provided. Perforations not less
than 5/16-inch in diameter or slots can be

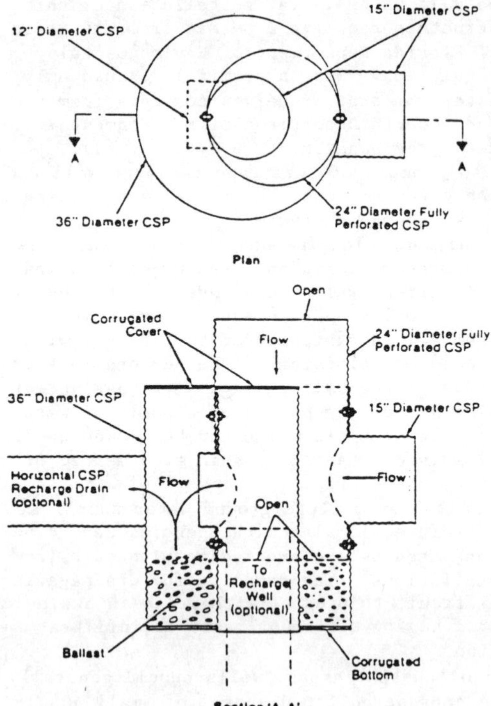

Figure 2 - Typical CSP dry well (3)

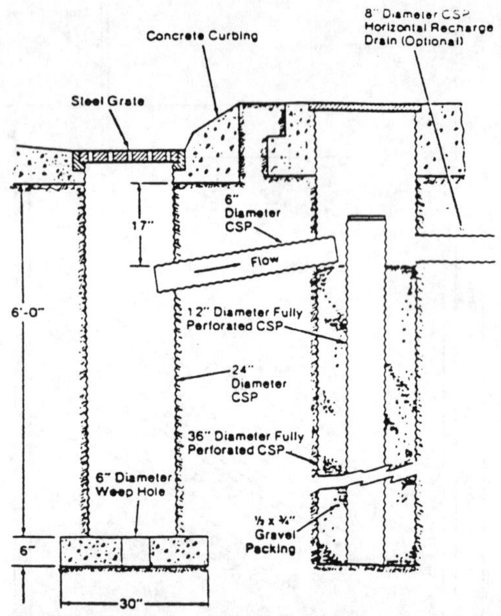

Figure 3 - CSP drop inlet and dry well (3)

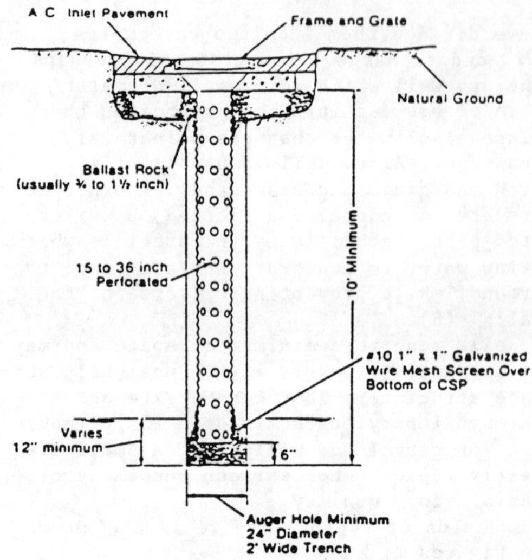

Figure 4 - Dry well (3)

used if they provide an opening area not less than 3.31 square inches per square foot of pipe surface. The photo in Figure 6 shows the inside of a metal pipe with perforations around the full periphery.

The liberal number of holes are to ensure free and rapid flow in and out of the pipe. The purpose of the large-sized pipes is to add to the total storage volume for stormwater and to reduce the quantity of expensive rock backfill.

Coarse gravel or other aggregate is used for backfilling the trench around, below and above the pipe so that part of the stormwater is temporarily stored in the voids of the backfill. Experiments made by the Dade County Florida Department of Public Works have indicated that 3/4-inch x 1-1/2 inch coarse gravel backfill with pipe systems having 3.31 square inches per square foot of perforations will provide pipe exfiltration rates which exceed the best infiltration rates of soils normally encountered in the field. (2).

The following is taken from the Dade County Florida Department of Public Works specifications covering pipe perforations and ballast rock for exfiltration trench construction:

35. Exfiltration Trench (Sections 443 and 514)(continued). .02 materials (continued). The number of perforations per linear foot of pipe are shown in the following table:

146

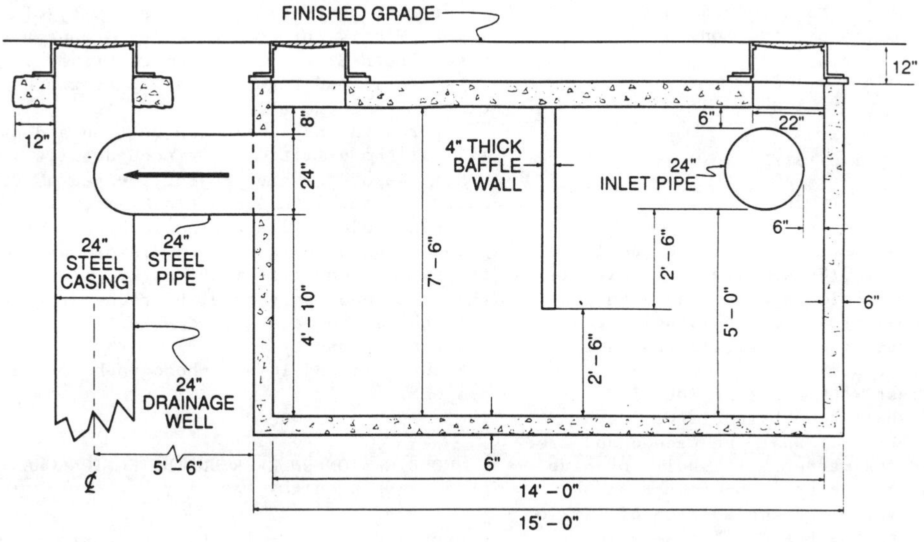

Figure 5 - Profile of grease interceptor and drainage well detail (3)

Figure 6 - Typical fully perforated corrugated steel pipe for
exfiltration trench construction

Pipe diameter   *3/8-inch diameter
in inches   perforations

15	100
18	120
24	160
30	200
36	240
84	560

*5/16-inch diameter holes may be utilized
in lieu of the 3/8-inch diameter holes if
the number of holes are increased to provide an equal cross sectional hole area.
The other requirements remain the same.

B. Ballast Rock:

All ballast rock shall be locally pro-
cured and shall be obtained from fresh
water sources. It shall be washed and free
of deleterious matter. It shall not have
more than 45 percent loss of section as
specified by the current edition of AASHTO
M63 governing the Los Angeles Abrasion
Test. It shall not show more than a 10
percent loss in 10 cycles as specified by
the current edition of AASHTO M63 governing the soundness test. The ballast rock
shall meet the gradation requirements as
specified by the current edition of AASHTO
M43 for size 24 (2-1/2 to 3/4-inch) or number 4 (1-1/2 to 3/4-inch). The County re-
serves the right to have sample tests made
of the material at selected intervals by
an approved laboratory at its expense.

Figures 7 and 7-A show a typical longitudinal section and cross-section for exfiltration trench drains used by Dade County
Florida.

Figure 8 is a photograph showing installation of 36-inch diameter perforated corrugated steel pipe, bituminous coated, along
the north side of Southwest 42nd Street
just east of 137th Avenue in Miami, Florida.

The criterion of retaining the first inch
of runoff has been used by federal and
state agencies. Studies by the Environmental Protection Agency have set the first
inch criteria as a guideline for water quality enhancement.

Although less strigent than total on-site
retention, it has been found that the first
inch of runoff contains nearly 100 percent
of all stormwater pollutants, while subsequent runoff is relatively free from pollutants. By sizing facilities to retain the
first inch from the entire drainage basin,
it is felt that maximum attention is given
to water quality at a time when water quality objectives cannot be fully attained.
(2).

Figure 9 and Figure 10 are the section
and plan views, respectively, using the design criteria of handling and treating the
first inch of runoff. This is a City of
Tampa, Florida project on Himes Avenue between Green Street and Columbus Drive.

Figure 11 and Figure 12 are photographs
of this same project.

Figures 13 and 14 are the section and profile of the exfiltration trench details for
a Home Depot private project located at Palmetto, NW and 57th Avenue in Miami, Florida.
In this project, 15-inch perforated corrugated steel pipe is used to dispose of the
stormwater runoff from the parking lot of
the Home Depot warehouse, by recharging it
into the subsurface ground water strata
under the property.

Figures 15 and 16 are photographs of this
project.

5 COMBINATION UNDERGROUND DETENTION CHAM-
BER AND RECHARGE FACILITY

Where land uses are intense, or where topography prohibits any appreciable surface
storage, a combination underground detention chamber and recharge facility should be
considered. Provision must be made to ensure that the overflow for storms exceeding
the design storm will not cause damage to
downstream properties. Other major design
considerations are provisions for cleaning
and prevention of debris from entering the
storage facility.

Figure 17 is a plan view of a combination
detention chamber and recharge facility.
There are 5 lines of 72-inch diameter
slotted corrugated steel pipe emptying
into a concrete control structure shown on
the upper left.

Figure 18 is a photograph of the 5 lines
of 72-inch diameter corrugated steel pipe
with the control structure shown in the
foreground.

Figure 19 is a photograph of the invert
of the 72-inch diameter corrugated steel
pipe showing the 6-inch slots in the inside valley of the corrugations. Slots are
on 10-inch centers; pipe is 5-inch x one-
inch corrugation profile.

Figure 20 is a photograph of the 5 lines
of 72-inch diameter corrugated steel pipe
taken from the end opposite from the control structure.

Figure 21 is an elevation view of the
control structure. Flow is from the 72-
inch diameter corrugated steel pipe (on
right hand side) into the structure. Note
the 1-foot x 1-foot low flow orfice in the
baffle wall.

Another variation from this previous combination arrangement is the combination retention basin and recharge facility. Figure 22 shows a plan view of a parking lot

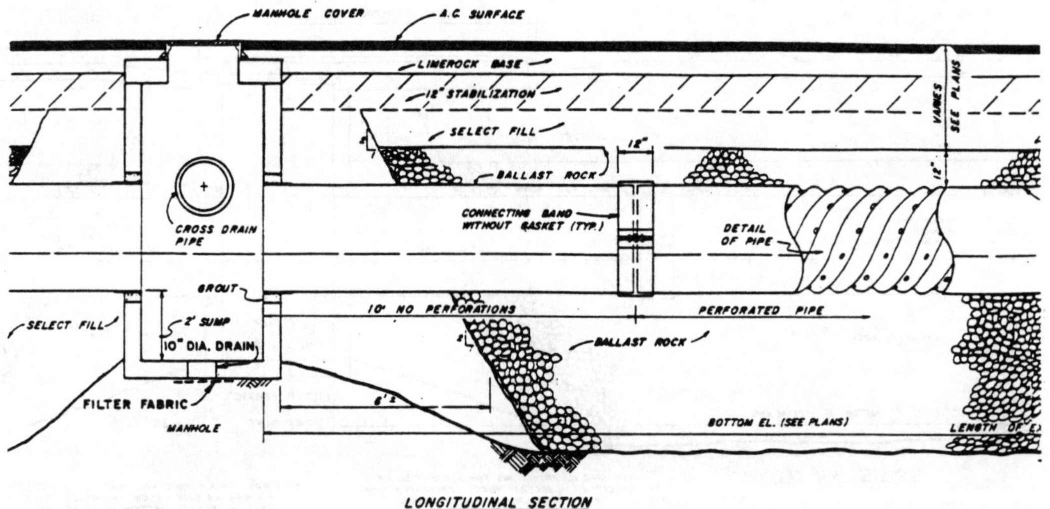

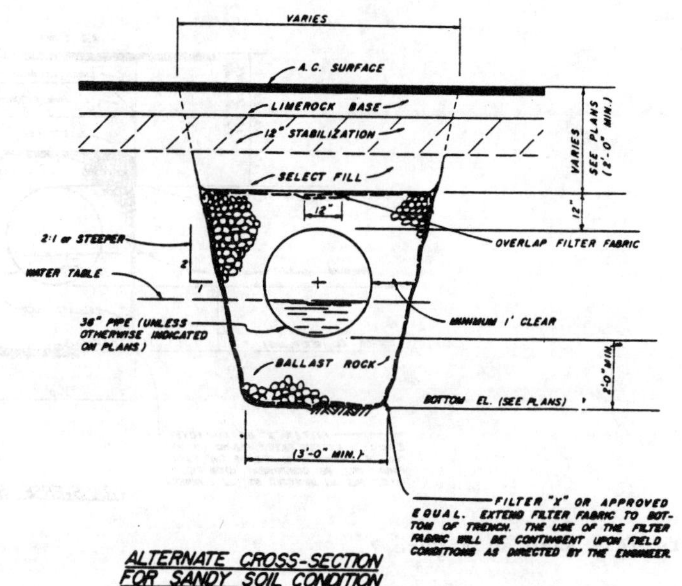

Figure 7 - These drawings taken from sheet 3-E titled "Detail Of Exfiltration Drain With Perforated Pipe." Dade County Florida Public Works Department, Highway Division. Date Rev. 11/89. (2)

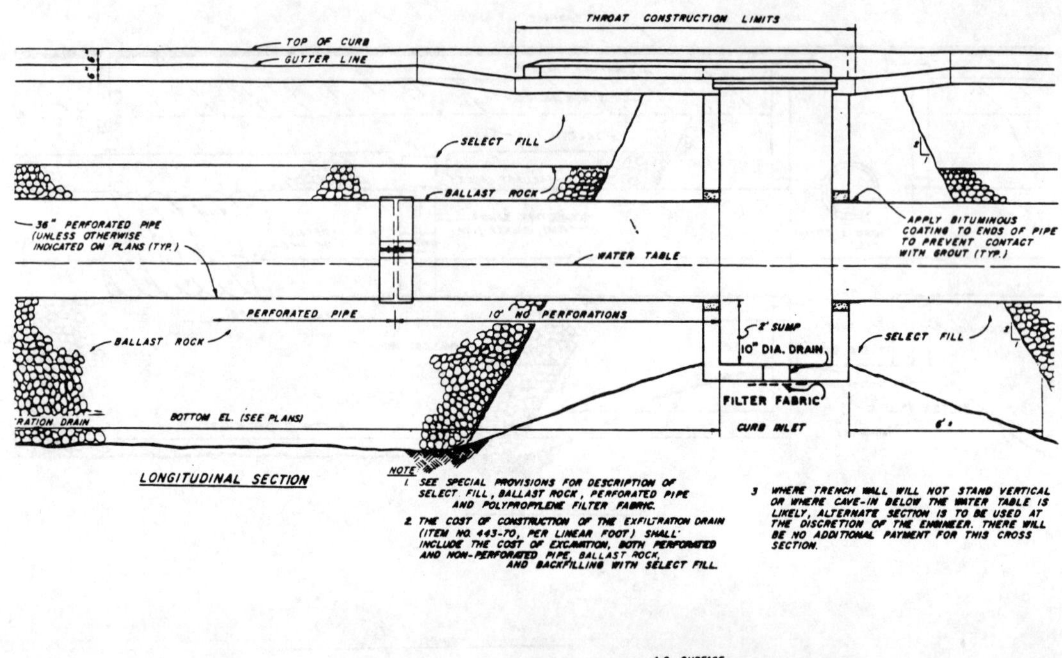

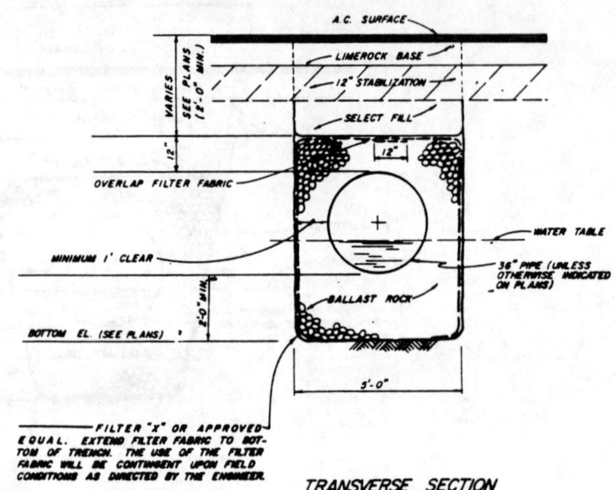

Figure 7A - Same as Figure 7.

Figure 8   Placing 36-inch diameter perforated corrugated steel pipe, bituminous-coated, in exfiltration trench in Miami, Florida

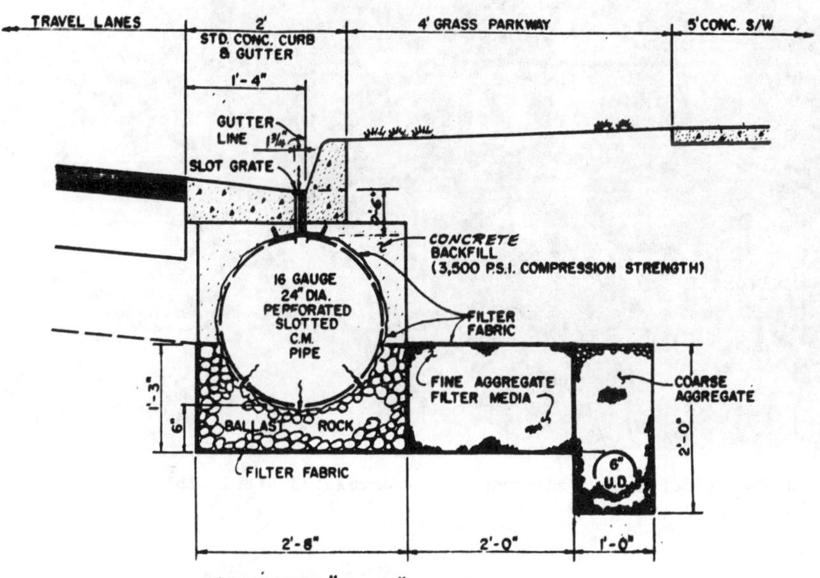

Figure 9 – Tampa, Florida Project – Himes Avenue between Green Street and Columbus Drive. Drawing courtesy DSA Group Tampa, Florida

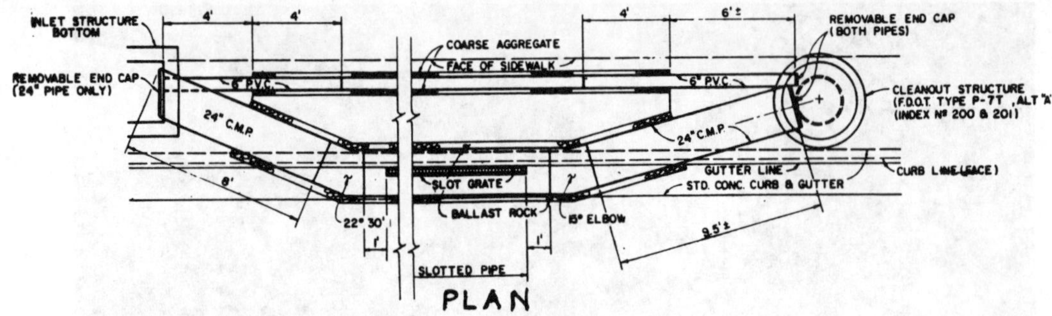

Figure 10 - Same as Figure 9

Figure 11 24-inch diameter perforated slotted drain corrugated steel pipe on Tampa, Florida Project along Himes Avenue

Figure 12   Same as Figure 11

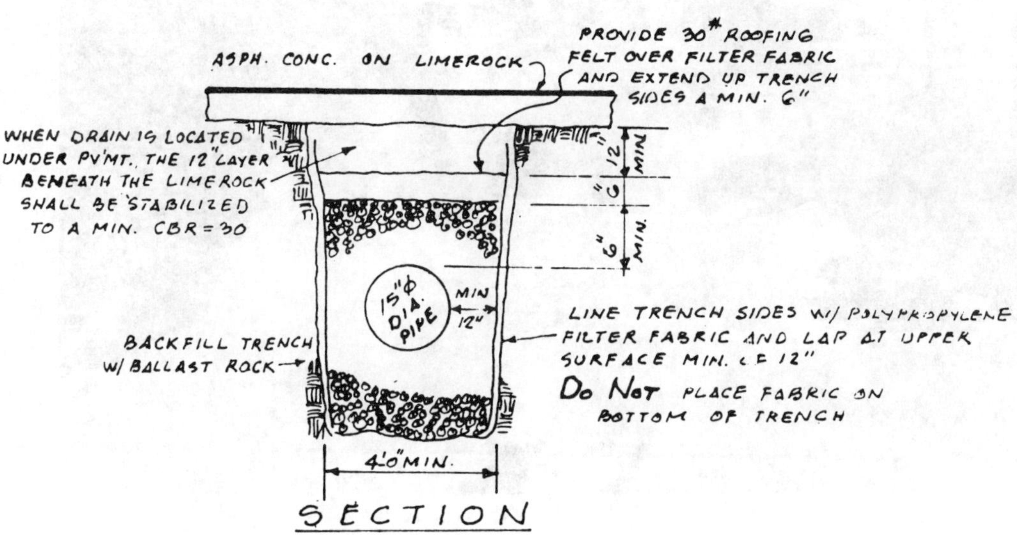

Figure 13 – Exfiltration trench construction details for
Home Depot Project at Palmetto, N. W. and
57th Avenue in Miami, Florida
Drawing courtesy Schwebke-Shiskin & Associates
Miami, Florida

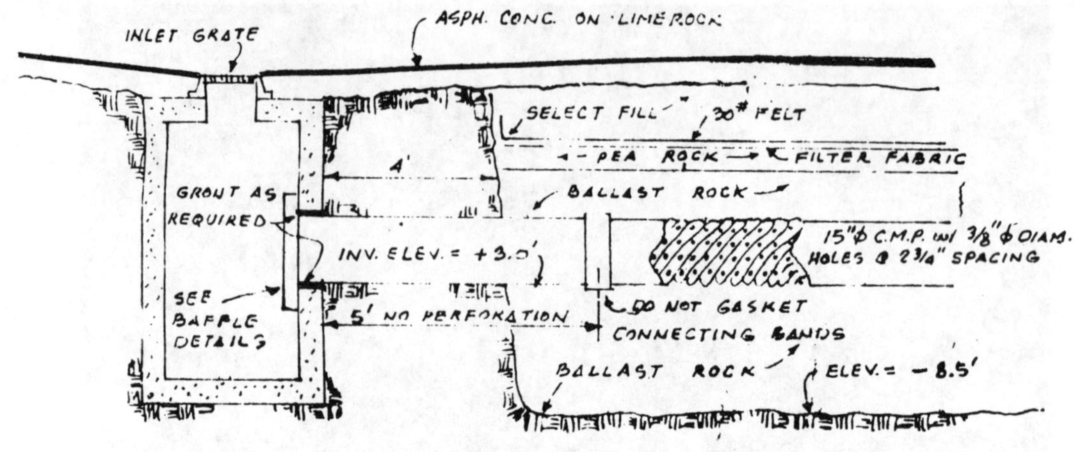

Figure 14 - Same as Figure 13

Figure 15   15-inch diameter perforated corrugated steel pipe used in exfiltration trench to dispose of parking lot runoff of Home Depot Project at Palmetto, NW and 57th Avenue in Miami, Florida

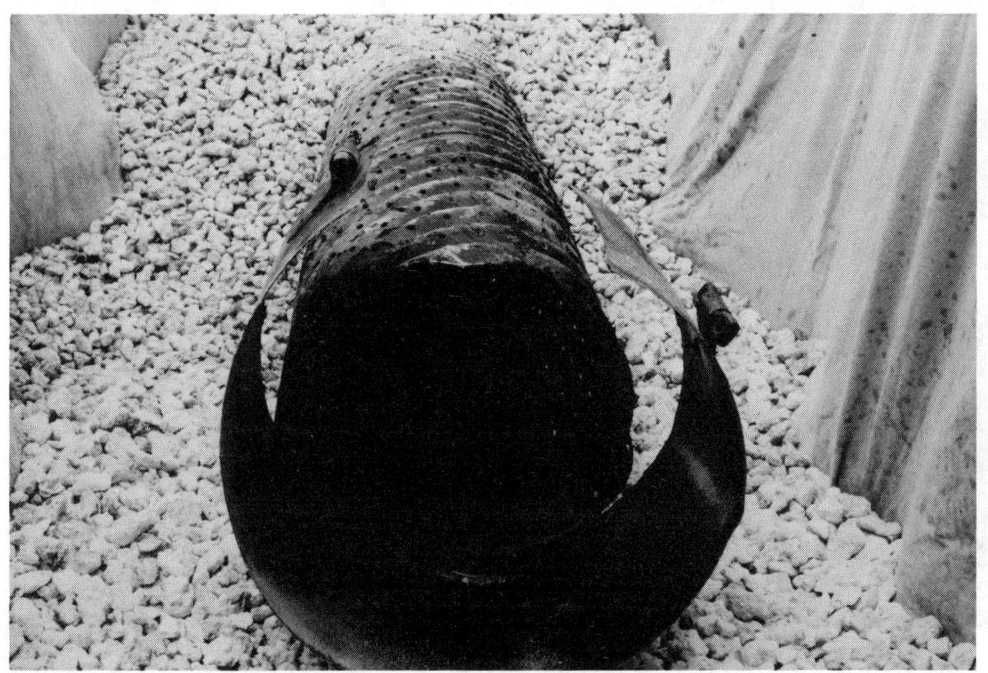

Figure 16   Same as Figure 15

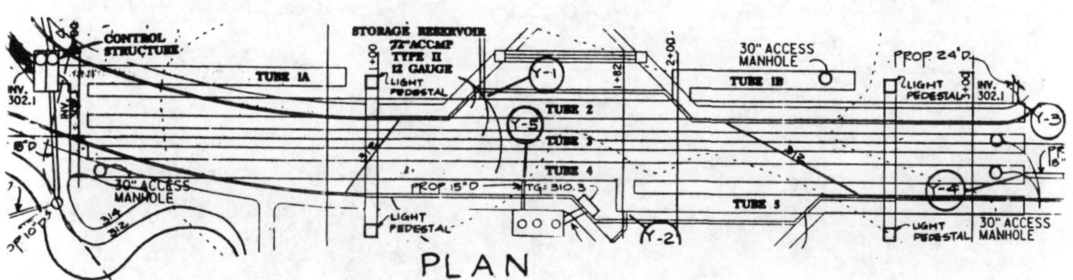

Figure 17 - 5 lines of 72-inch slotted corrugated steel pipe emptying into control structure
at upper left, located in Baltimore County, Maryland.
Drawing curtesy of Daft McCune-
Walker, Inc., Towson, Maryland

Figure 18   72-inch diameter slotted corrugated steel pipe with control structure shown in foreground. Baltimore, Maryland

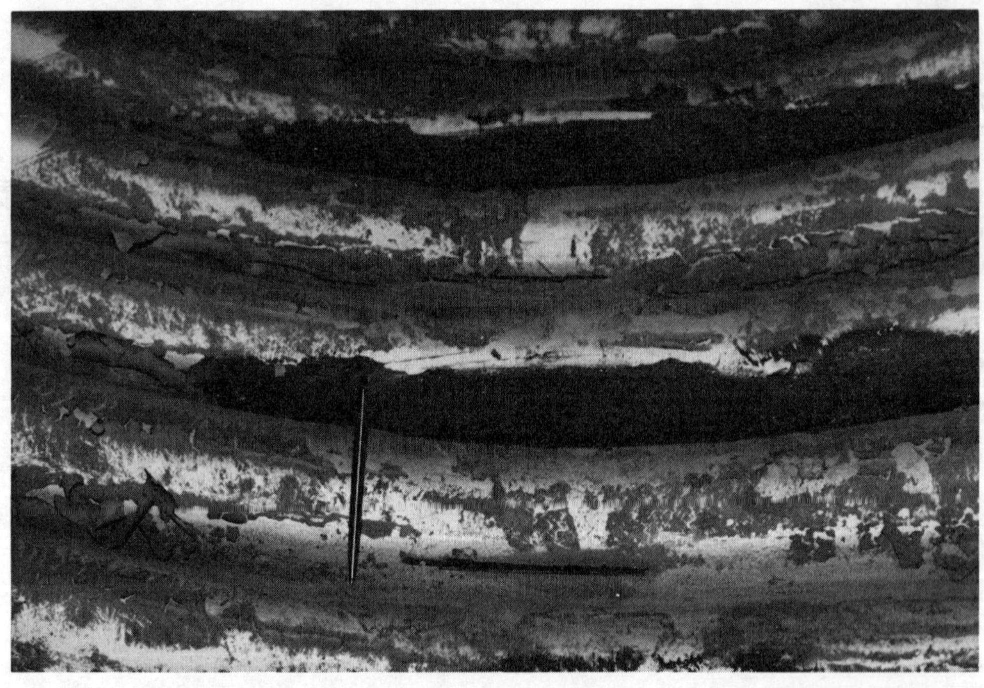

Figure 19   Close-up of invert of 72-inch corrugated steel pipe with 6-inch slot on inside valley of corrugations. Slots are on 10-inch center; pipe is 5"x1" corrugation profile. Baltimore County, Maryland

Figure 20 5 lines of 72-inch diameter slotted corrugated steel pipe. This view from opposite end of control structure. Baltimore County, Maryland

Figure 20-A 18 lines of 48-inch diameter fully perforated corrugated steel pipe used as a combination detention chamber and recharge system. A Toys "R" Us store and parking lot in Ocala, Florida

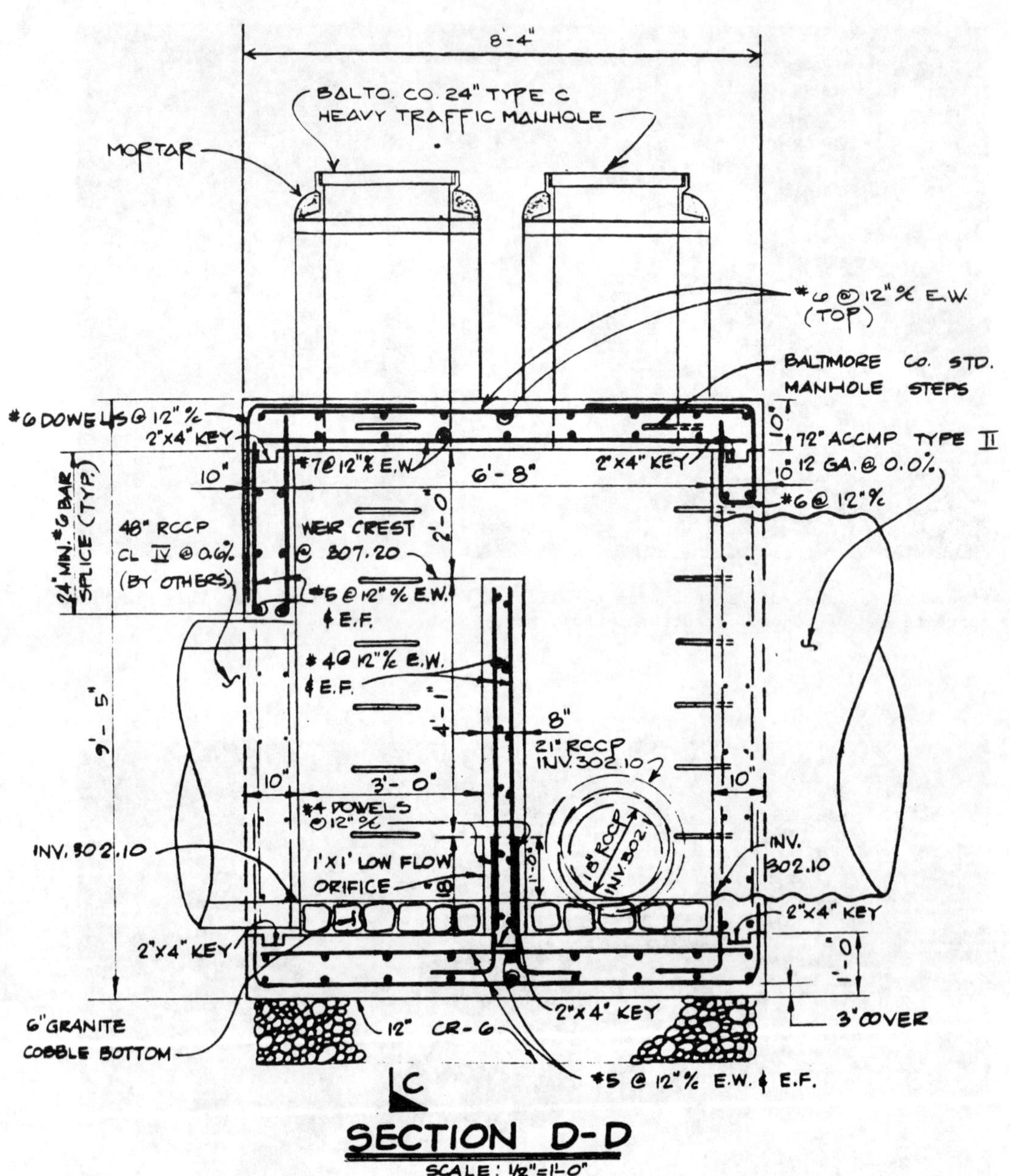

Figure 21 - This is an elevation view of the control structure. The
72-inch corrugated steel pipe empties into the structure
from right hand side. Baltimore County, MD
Drawing courtesy Daft McCune-Walker, Inc., Towson, MD

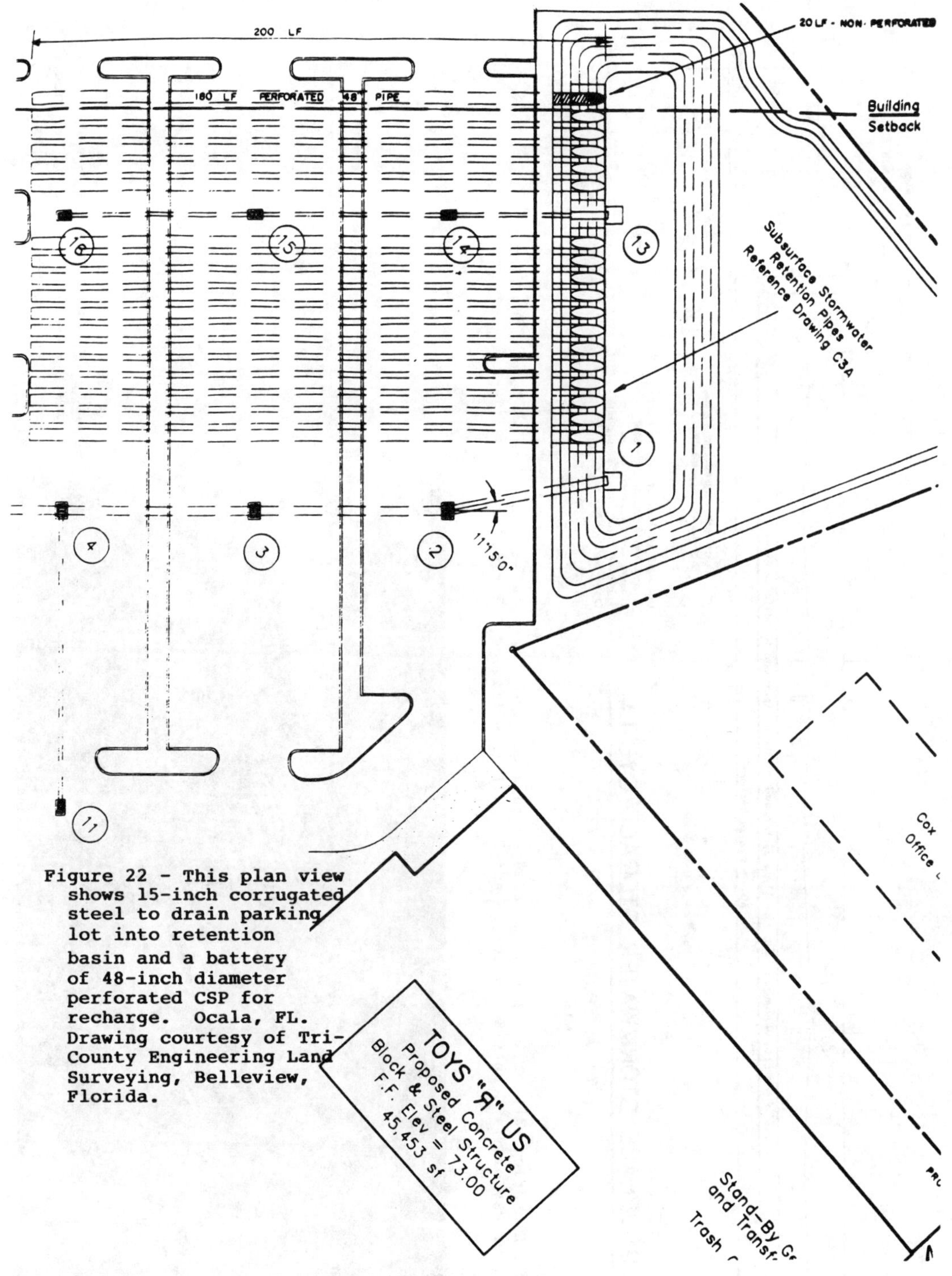

Figure 22 - This plan view shows 15-inch corrugated steel to drain parking lot into retention basin and a battery of 48-inch diameter perforated CSP for recharge. Ocala, FL. Drawing courtesy of Tri-County Engineering Land Surveying, Belleview, Florida.

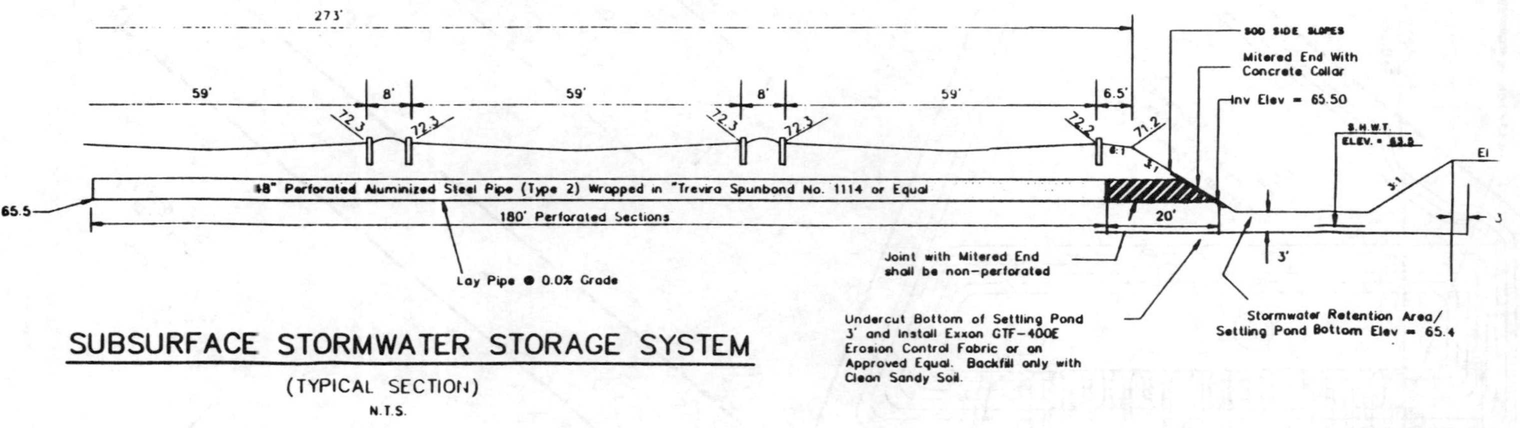

Figure 23 Elevation view of 48-inch perforated corrugated steel pipe and section view of retention basin, Ocala, Florida. Drawing courtesy of Tri-County Engineering-Land Surveying, Belleview, Florida

Figure 24 Retention basin and beveled ends of 18 lines of 48-inch perforated corrugated steel pipe, Ocala, Florida

Figure 25   Catch basin and 15-inch corrugated steel pipe for draining parking lot shown at right.   A Toys "R" Us store and parking lot in Ocala, Florida

Figure 26   Partial view of the 18 lines of 48-inch diameter perforated corrugated steel pipe used as a recharge system from the retention basin.   A Toys "R" Us store and parking lot in Ocala, Florida

for Toys "R" Us facility in Ocala, Florida; 15-inch corrugated steel pipe is used
for carrying runoff from catch basins 16,
15, 14, 4, 3 and 2 into a retention basin
on the right. When the water level in the
retention basin rises to the elevation of
the battery of 18 lines of 48-inch fully
perforated corrugated steel pipe, it flows
into them and is recharged into the soil
strata under the parking lot.

Figure 23 shows an elevation view of the
48-inch perforated corrugated steel pipe
and a section view of the retention basin.

Figure 24 is a photograph of the retention basin and the 18 lines of 48-inch
perforated corrugated steel pipe.

Figure 25 is a photograph showing one of
the catch basins on the right and the 15-
inch diameter corrugated steel pipe which
carries the parking lot runoff to the retention basin. Also shown is a portion of
the 18 lines of 48-inch diameter fully perforated corrugated steel pipe.

Figure 26 is a photograph showing the 48-
inch diameter fully perforated corrugated
steel pipe used for recharging being installed.

6 SUMMARY

Stormwater management planning has been
practiced in some areas of the United
States for the past two decades and all of
the infiltration/recharge systems heretofore mentioned in this paper have been
used in this planning at some time or the
other. However, these systems must be site
adaptable--that is to say, what works well
in one place may not be the best in another.
Down through the years, because of the public's pressure and demand for a cleaner environment, federal, state, county, town-
ship, municipal, regional and district en-
tities have tightened their flood control
and drainage programs to the point that
some of these systems are not feasible.
The impact from this clean environment
movement will be felt even more when Environmental Protection Agency regulations
aimed at governing the discharge of stormwater from industrial facilities, to be en-
acted in October, 1990, are implemented. It
would seem the combination underground detention chamber and recharge facility
would be the logical system for industry
to use to comply with this latest regulation. It would solve the problem in most
instances, plus the area in the vicinity
of the facility reaps the benefits of recharging the stormwater back to the underground strata or aquifer.

REFERENCES AND BIBLIOGRAPHY

1. Curtis, R. F."Water And Our Future,
VISaVIS Magazine, Vol. 4, No. 4, East West
Network, Inc., New York, NY, April 1990.
2. Dade County, Florida Department of En-
vironmental Resources Management, "Design
Of Drainage Structures: An Updated Policy
For The Design Of Storm Runoff Drainage
Structures," December 1980.
3. "Modern Sewer Design," American Iron
& Steel Institute, Washington, D. C., 2nd
Ed., 1990.
4. ASCE Proceedings Of The International
Symposium, "Artifical Recharge Of Ground
Water," Anaheim, CA, August 23-27, 1988.
5. American Public Works Association, "Urban Stormwater Management," Special Report
No. 49, 1981.
6. "Underground Disposal Of Stormwater Run-
off," Design Guidelines Manual, U. S. Dept.
of Transportation, FHWA, February 1980.

Structural Performance of Flexible Pipes, Sargand, Mitchell & Hurd (eds) © 1990 Balkema, Rotterdam. ISBN 90 6191 165 6

Predicting the behavior of buried pipes using a mixed finite element approach

Musharraf Zaman & Joakim G. Laguros
University of Oklahoma, Norman, Okla., USA

ABSTRACT: A mixed finite element procedure is presented in this paper for stress-deformation analysis of buried pipes taking into consideration the behavior of the interface between the pipe and the surrounding soil. Both relative displacements and stresses at the interface are treated as primary variables. A Mohr-Coulomb type slip function is used to characterize the slip behavior of interface. The finite element procedure was used to analyze two problems. In the first problem, the distribution of stresses in stiff soil induced by a deeply buried (flexible) pipe was compared with the analytical solutions. Influence of interface conditions (smooth, frictional, bonded) was studied in the second problem and the results were compared with existing solutions and satisfactory agreement was observed. The mixed finite element procedure presented here can be used to accurately analyze stress-deformation response of buried pipes of practical importance.

1 INTRODUCTION

Buried pipes are used extensively as sewer lines, roadway culverts and in the transportation of oil and natural gas. In contrast to above ground pipelines, the structural design of buried pipes should take into account the mutual interaction between the pipe and the surrounding soil (Selvadurai, 1985). The soil-structure interactions involving buried pipes can be induced by various factors such as the pipe geometry and material properties, the differences in the properties between the natural soil and the compacted backfill, the condition of the soil-pipe interface and the nature and magnitude of loads acting on the pipe. An analysis of the complete soil-pipe interaction including various contributing factors involved is an extremely difficult task. Therefore, usually simplified assumptions are made to make such analyses mathematically tractable. The advent of high speed, electronic computers has made the use of numerical techniques for analysis of buried pipes attractive and cost effective, and of these the Finite Element Method (FEM) is probably most versatile.

This paper presents a finite element-based analysis procedure that accounts for the interaction between the pipe and the surrounding soil medium using a "mixed interface element." Plane strain idealization is adopted. A mixed variational principle based on the kinematic constraints associated with various deformation modes is utilized to generate the stiffness matrix of the element. An iterative algorithm, in which both displacements and tractions are treated as independent variables, is used to simulate the deformation modes. The mixed-FEM is used to analyze the flexural behavior of a buried pipe. The pipe material and the soil medium are treated as elastic and the interface elements are assumed to exhibit nonlinearity. A parametric study is conducted to identify the influence of interface conditions on overall response of the pipe.

2 FINITE ELEMENT ANALYSIS

In the finite element analysis adopted in this study, the displacement FEM formulation is utilized to model the buried pipe and the surrounding soil medium, while the mixed formulation is used for the interface elements. Plane strain idealization is employed for simplicity. Such a representation is particularly suitable for a long pipeline having homogeneous field conditions. Eight-noded isoparametric elements are used for discretization of the pipe and

the soil medium. Alternatively, curved
shell elements can be used to idealize the
pipe. Derivation of the stiffness matrix
for eight-noded isoparametric elements is
given in the literature (e.g., Cook, 1981;
Zienkiewicz, 1977).

3 MIXED-INTERFACE ELEMENT

No attempt is made here to review the extensive literature available on modeling of
contact problems. Comprehensive reviews of
the existing interface models can be found
in various publications (e.g., Desai et al.
(1984), Zaman (1982) and Hohberg and Bachmann (1988)). In the present study, incor-
poration of various deformation modes (i.e.,
contact, slip, debonding, rebonding) in the
interface element is based on the traction
and displacement values at the interface.
These traction and displacement values must
satisfy certain constraints in order that
the deformation be physically admissible.
For example, when opening or debonding occurs at an interface, neither normal nor
shear traction can be transmitted through
the interface. In this situation, both the
normal stress, σ_{nn}^{o} and the shear stress,
τ^{o} must be zero for the debonded region.
Similarly, when an interface element is in
stick or contact mode, the relative displacements, u_{r}^{o}, for that element must van-
ish. These and other traction and displacement criteria can be easily satis-fied if
both the tractions and displacements are
the primary unknowns for an interface element. Formulation of the mixed interface
element adopted here is essentially based
on this concept (Fig. 1). The element has
nine nodes and zero thickness. Six nodes
(shown by o) associated with the surrounding solid elements have displacements as
primary unknowns, while the remaining three
nodes (shown by x) have tractions as primary
unknowns. A variational principle is utilized to generate stiffness matrix for the
element.

For illustration, assume that the interaction between body 1 and body 2 (Fig. 1)
is characterized by the total contact region, consisting of stick region, S_{st}, slip
region, $S_{s\ell}$ and separation region, S_{sp}.
For a given load level, a point on the in-
terbody boundary region S_{t} falls into or
remains in one of the three modes, namely,
stick mode, slip mode or separation mode.

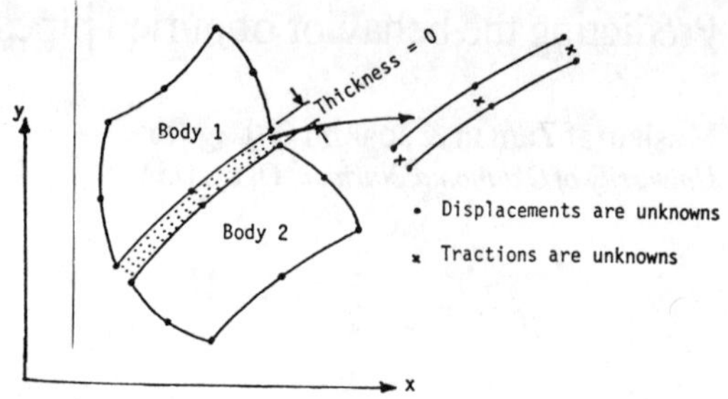

Figure 1. Mixed Interface Element Used

For stick and slip modes, the normal
traction, t_{n}^{o} is compressive and the total
normal relative displacement, u_{n}^{o} is zero.
That is

$$t_{n}^{o} \geq 0 \tag{1a}$$

and

$$u_{n}^{o} = 0 \text{ on } S_{st} \text{ and } S_{s\ell} \tag{1b}$$

For slip mode, the slip function

$$F = c_{a} + \sigma_{nn}^{o} \tan \delta - |\tau^{o}| \tag{2a}$$

and

$$(s_{i} - n_{i} \tan \delta) \Delta t_{i} = 0 \tag{2b}$$

where c_{a} and δ are the apparent cohesion
and friction angle of the interface,
respectively, σ_{nn}^{o} is the total normal
stress and τ^{o} is the total shear stress,
s_{i} and n_{i} are unit vectors along the direction of tangential and normal tractions,
respectively, and Δt_{i} represents incremental traction. For separation mode, the
total traction t_{i}^{o} is equal to zero. If the
point remains in separation mode, the incremental traction Δt_{i} is also zero. That is

$$t_{i}^{o} = 0 \tag{3a}$$

and

$$\Delta t_{i} = 0 \text{ on } S_{sp} \tag{3b}$$

The conditions on incremental relative displacement Δu_{r} for stick and slip modes can
be combined into a single expression

164

$$\Delta u_i = \lambda' \, s_i, \text{ on } S_{st} \text{ and } S_{s\ell} \qquad (4)$$

where $\lambda' = 0$ on S_{st} and S_{sp}; $\lambda' > 0$ on $S_{s\ell}$.

The constraint conditions related to various deformation modes can be included in the finite element formulation by making use of a modified variational principle. A variational principle similar to that used by Urzua et al. (1977) for analysis of bolted connections is used here and is based on the functional $I\,(\Delta u_i, \Delta t_i, \lambda')$.

$$I = \iint_{S_t} \Delta t_i [\Delta u_i - \lambda' \, s_i]\, dS \qquad (5)$$

The variables in Eq. (5) satisfy the following forced boundary conditions,

$$\Delta u_i = \overline{\Delta u}_i \text{ on } S_u; \; \Delta t_i = 0 \text{ on } S_{sp};$$

$$\lambda' = 0 \text{ on } S_{st} \text{ and } S_{sp}; \text{ and } \lambda' > 0 \text{ on } S_{s\ell}.$$

The incremental relative displacements Δu_i in Eq. (5) can be expressed as

$$\Delta u_i = N^m \Delta u_i^m \qquad (6)$$

where N^m are the interpolation or shape functions and Δu_i^m are the relative displacements evaluated at nodal point m. Similarly, the traction Δt_i can be expressed in terms of interpolation function as

$$\Delta t_i = N^m \Delta t_i^m \qquad (7)$$

where Δt_i^m is the incremental traction evaluated at node m. Also, the incremental shear traction $(\Delta t_i . s_i)$ can be written as

$$\Delta t_i \, s_i = N^m \Delta t_i^m s_i^{(m)} \qquad (8)$$

where $s_i^{(m)}$. Δt_i^m represents the slip constraints evaluated at node m. The superscript (m) does not inter in summation convention.

Substitution of Eqs. (6), (7) and (8) into Eq. (5) and simplification lead to

$$I = \Delta t_i^m \Delta u_i^n \left[\iint_{S_t} N^m N^n \, dS \right] - s_i^{(m)}$$

$$\Delta t_i^m \left[\iint_{S_t} N^m \lambda' \, dS \right] \qquad (9)$$

or

$$I = \Delta t_i^m \Delta u_i^n C_{mn} - s_i^{(m)} \Delta t_i^m \Lambda^m \qquad (10)$$

where

$$C_{mn} = \iint_{S_t} N^m N^n \, dS$$

are the interface stiffness coefficients, and Λ^m are the Lagrangian multipliers for those nodes which are in slip modes.

The sum of the interface functional, I_T, for all the interface elements can be expressed as

$$I_T = \{\Delta t_i\} \, [C_{mn}] \, \{\Delta u_i\} - \{\Lambda\}^T \, [S_i] \, \{\Delta t_i\}^T \qquad (11)$$

where the superscript T denotes transpose of a matrix. The matrices in Eq. (11) are obtained by appropriately adding the contributions of all the interface elements. The equilibrium equations can now be obtained by giving an independent variation to each of the variables in the functional I_i (viz, Δu_i, Δt_i, and Λ). Thus,

$$\delta\,(I_T) = \{\delta\,\Delta t_i\} \, [C_{mn}] \, \{\Delta u_i\}$$
$$+ \{\delta\,\Delta u_i\} \, [C_{mn}] \, \{\Delta t_i\} - \{\delta\,\Lambda\}^T \, [S_i]^T \, \{\Delta t_i\}$$
$$- \{\delta\,\Delta t_i\} \, [S_i] \, \{\Lambda\} \qquad (12)$$

Note that since both tractions and displacements are treated as primary unknowns, the global stiffness matrix will contain zero terms on the main diagonal and may not be positive definite. An equation solver capable of handling this situation may be required. However, the present study has shown that if the number of interface elements is fewer (or significantly less) than the number of pipe and soil elements, and if the node numbers are marked in such a way that the interface nodes do not appear in the beginning, then an ordinary equation solver based on Gaussian elimination is found adequate. Further details , including the iterative procedure used for the simulation of deformation modes, are given elsewhere (Zaman, 1982).

4 ANALYSIS OF BURIED PIPES

The mixed-FEM discussed in the preceding section was used to analyze the flexural behavior of buried pipes. In the first problem considered here, distribution of stresses within the soil medium induced by a deeply buried circular pipe of diameter D is analyzed (Fig. 2). The soil medium is assumed to be stiff, compared with the rigidity of the embedded pipe, so that the problem can be treated as a

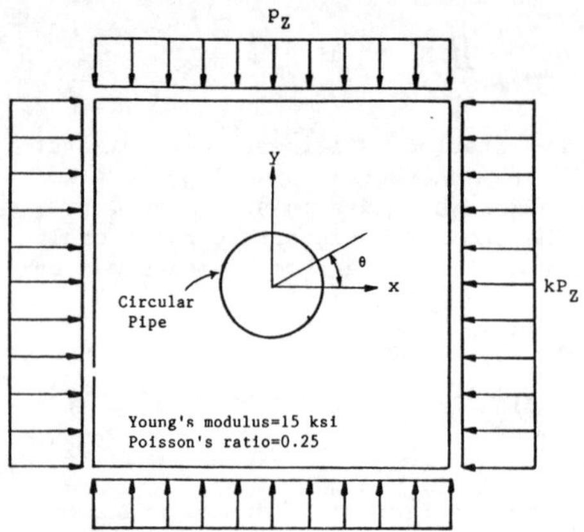

Figure 2. Finite Element Idealization of a Deeply Embedded Circular Pipe in Homogeneous Stiff Soil

circular tunnel problem for which analytical solutions are available (Brown and Hoek, 1980). The reason for solving this problem is to verify the finite element results.

Figure 2 shows the idealized loading, due to overburden, considered in the analysis. The intensity of the vertical load acting at the top and the bottom of the idealized domain is designated by P_z.

This assumption, of course, is applicable to deeply buried pipes only where the weight of the soil for the idealized domain is negligible compared to the overburden P_z.

The intensity of the lateral load is given by $K\,P_z$, K being the earth pressure coefficient. In view of symmetry, only one-quarter of the domain is analyzed.

Figure 3 shows the comparison of the circumferential stress, σ_θ, along the periphery of the pipe. Some minor differences between the finite element results and the theoretical values are observed. The possible reason for this difference is that the analytical solution represents the circumferential stresses at the free surface of the excavation while finite element results are obtained at integration points which are located slightly inside the domain. Figures 4 and 5 show the variation of circumferential stresses along the spring line and crown line of the pipe, respectively. Overall, an excellent agreement is obtained. For $\theta = 0$, the maximum stress concentration occurs at the crown. A somewhat different behavior is observed for $\theta = 90^\circ$ degrees.

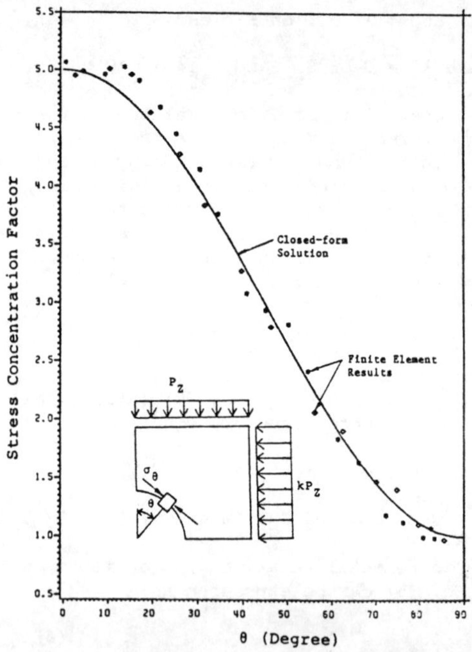

Figure 3. Comparison Between Closed-Form and Finite Element Results for σ_θ Along Excavation Boundary (Homogeneous Medium, k=2.0).

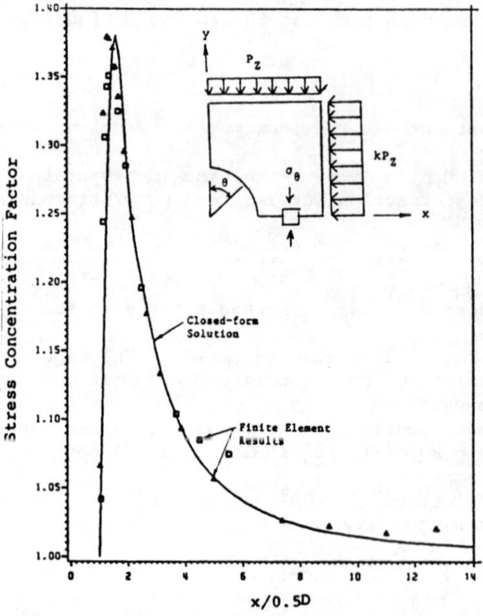

Figure 4. Comparison Between Closed-Form and Finite Element Results for σ_θ Along Spring Line (Homogeneous Medium, k=2.0).

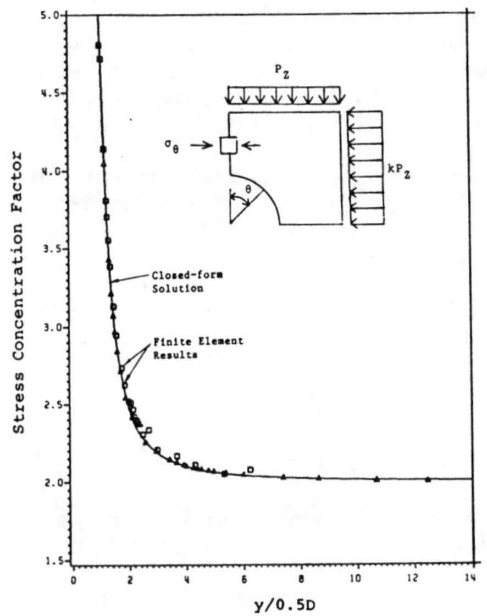

Figure 5. Comparison Between Closed-Form
and Finite Element Results Along Crown Line
(Homogeneous Medium, k-2.0).

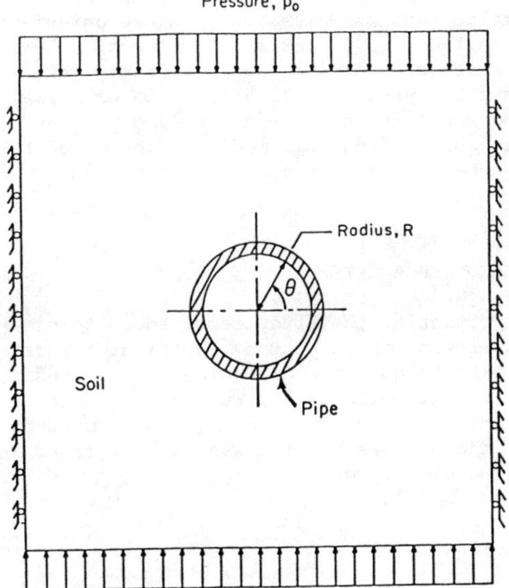

Figure 6. Buried Pipe Problem Including
Interface.

4.1 Analysis of buried pipe including interface friction

Figure 6 shows two-dimensional idealization
of a pipe-soil system analyzed by Katona

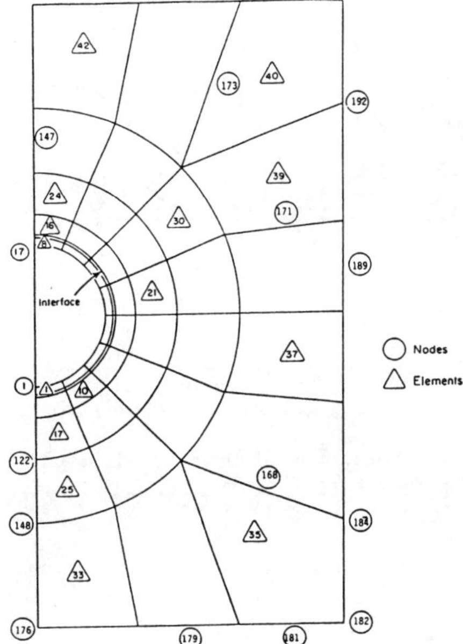

Figure 7. Finite Element Mesh Used in
Simulation of Buried Pipe Problem

(1981) using a "contact-friction" interface
element and by Desai et al. (1984) using
the "thin-layer element". Using the same
material and interface properties as used
by Desai et al. (1984) and the mesh shown
in Fig. 7, the buried pipe problem is
solved using the mixed finite element.
Three cases of interface friction are considered: tan δ = 0.001, 2.0 and 0.25 to
represent approximately frictionless slip,
stick, and frictional slip conditions, respectively. The value of tan δ is used to
define the slip function according to the
Mohr-Coulomb criterion. Typical results
obtained from this analysis are presented
in Figs. 8-10.

Figure 8 shows comparison of computed
$\dfrac{\sigma_{rr}}{P_o}$ vs θ (from springline to crown) curves
with the thin-layer element results (Desai
et al., 1984) and with the closed-form solutions. For the bonded and frictionless
cases the computed results are in very close
agreement with the exact solutions. In the
crown region (tan δ = 0.25), the computed
values are higher than the predictions obtained from the idealization of interface
by the thin-layer element. Figure 9 shows
a similar comparison for τ/P_o versus θ.

167

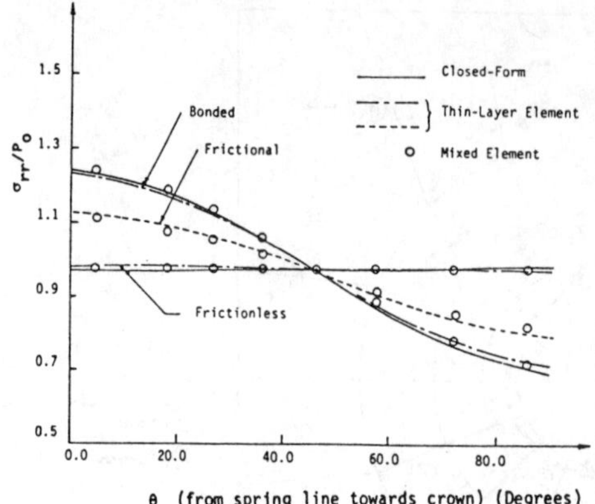

Figure 8. Variation of Nondimensional Radial Stress With Θ.

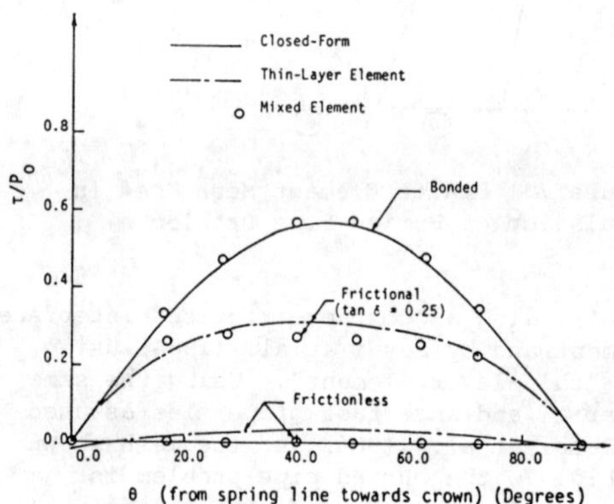

Figure 9. Variation of Nondimensional Shear Stress With Θ

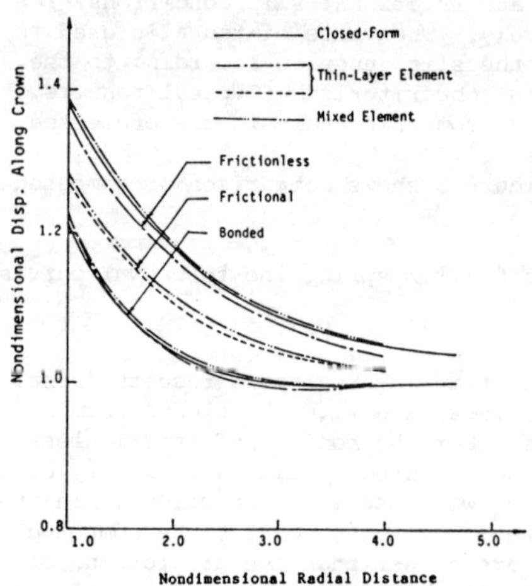

Figure 10. Variation of Nondimensional Displacement Along Crown With Radial Distance.

For frictionless and bonded cases the present computations compare very closely with the exact solutions. For the frictional case (tan δ = 0.25) the computed results are lower than the predictions of the thin-layer element and closer to Katona's results; however, the difference is not significant. Comparison of radial displacements, u_r is shown in Fig. 10. It is seen that for the frictionless and bonded cases the computed results are very close to the analytical values. As noticed previously, for the frictional case the computed results are slightly higher than the corresponding thin element results.

5 CONCLUDING REMARKS

A mixed finite element procedure is presented in this paper for stress-deformation analysis of buried pipes including the behavior of the interface between the pipe and the surrounding soil. Both relative displacements and stresses at the interface are treated as primary variables. A Mohr-Coulomb type slip function is used to characterize the slip behavior of interface.

The mixed FEM procedure applied to the flexural analysis of buried pipes yields results which are in close agreement with those of the existing methods. The interface conditions can noticeably influence the values and distribution of normal and shear stresses in the buried pipe and the surrounding soil medium. The mixed FEM can be effectively used to analyze such problems.

ACKNOWLEDGEMENTS

A portion of the study reported in this paper was carried out by the first author under the guidance of Professor C.S. Desai, University of Arizona, Tucson. Assistance of Mr. S.L. Kuang, former graduate student of the University of Oklahoma, is gratefully acknowledged.

REFERENCES

Brown, E.T. and Hoek, E. 1980. Underground Excavations of Rock. Institution of Mining and Metallurgy. London.

Cook, R.D. 1981. Concepts and Applications of Finite Element Analysis. Second Ed. New York: John Wiley.

Desai, C.S., M.M. Zaman, J.G. Lightner & H.J. Siriwardane 1984. Thin-Layer Element for Interfaces and Joints. Int. J. for Numerical and Analytical Met. in Geomech. 8:1:19-43.

Hohberg, J.M. & H. Bachmann 1988. A Macro
Joint Element for the Nonlinear Analysis
of Arch Dams. Proc. 6th Intl. Conf. Num.
Met. in Geomech. Innsbruck, Austria.
2:829-834.

Katona, M.G. 1981. A Simple Contact-Friction
Interface Element With Application to Buried Culvert. Proc. Implementation of Comp.
Procedures and Stress-Strain Laws in Geotech. Eng. Chicago. p. 45-63.

Kuang, S.L. 1983. Evaluation of Interface
Behavior in Soil-Structure Interaction.
M.S. Thesis. Univ. of Oklahoma. Norman.

Selvadurai, A.P.S. 1985. Numerical Simulation of Soil-Pipeline Interaction in a
Ground Subsidence Zone. Proc. Advances
in Underground Pipeline Eng. ASCE. Univ.
of Wisconsin-Madison. p. 311-319.

Urzua, J.L. & O.A. Pecknold 1977. Analyses
of Frictional Contact Problems Using an
Interface Element. Proc. Symp. on Appl.
of Comput. Meth. in Engg. Aug. 23-26.
Los Angeles, CA.

Zaman, M.M. 1982. Influence of Interface
Behavior in Dynamic Soil-Structure Inter-
action Problems:, Ph.D. Dissertation.
Univ. of Arizona. Tucson.

Zaman, M.M. & C.S. Desai. 1983. Models for
Sliding and Separation at Interfaces Un-
der Static and Cyclic Loading. Proc. Int.
Conf. on Constitutive Laws for Engineering
Materials: Theory and Application. Univ.
of Arizona. Tucson. p. 383-390.

Zienkiewicz, O.C. 1977. The Finite Element
Method. Third Ed. London: McGraw-Hill.

Structural Performance of Flexible Pipes, Sargand, Mitchell & Hurd (eds) © 1990 Balkema, Rotterdam. ISBN 90 6191 165 6

Author index